CAMBRIDGE LIBRARY COLLECTION

Books of enduring scholarly value

Religion

For centuries, scripture and theology were the focus of prof
scholarship and publishing, dominated in the English-speaking world by the work
of Protestant Christians. Enlightenment philosophy and science, anthropology,
ethnology and the colonial experience all brought new perspectives, lively debates
and heated controversies to the study of religion and its role in the world, many
of which continue to this day. This series explores the editing and interpretation
of religious texts, the history of religious ideas and institutions, and not least the
encounter between religion and science.

From Comte to Benjamin Kidd

Robert Mackintosh (1858-1933), a professor at the Congregationalist Lancashire
Independent College, traces the influence of biology and evolutionary thought on
the study of human ethics and society during the second half of the nineteenth
century in this 1899 book. He begins with Comte's founding of sociology, and
continues with the renewed appeal to biology for the understanding of human
affairs found in the work of Darwin, Spencer and their circle. He then looks at
Benjamin Kidd's Social Evolution, published in 1894 (and also reissued in this
series). Fifty years after Comte, Kidd argued that sociology required further
grounding by a new recourse to biology Mackintosh supported Kidd's view. If
biological clues are to afford guidance for human conduct, Mackintosh contended,
they must be supplemented by a clearer moral and religious vision, and in
philosophy by some scheme of metaphysical evolutionism. His work marks a
transition from Darwinism to a new Hegelianism.

Cambridge University Press has long been a pioneer in the reissuing of out-of-print titles from its own backlist, producing digital reprints of books that are still sought after by scholars and students but could not be reprinted economically using traditional technology. The Cambridge Library Collection extends this activity to a wider range of books which are still of importance to researchers and professionals, either for the source material they contain, or as landmarks in the history of their academic discipline.

Drawing from the world-renowned collections in the Cambridge University Library, and guided by the advice of experts in each subject area, Cambridge University Press is using state-of-the-art scanning machines in its own Printing House to capture the content of each book selected for inclusion. The files are processed to give a consistently clear, crisp image, and the books finished to the high quality standard for which the Press is recognised around the world. The latest print-on-demand technology ensures that the books will remain available indefinitely, and that orders for single or multiple copies can quickly be supplied.

The Cambridge Library Collection will bring back to life books of enduring scholarly value (including out-of-copyright works originally issued by other publishers) across a wide range of disciplines in the humanities and social sciences and in science and technology.

From Comte to Benjamin Kidd

The Appeal to Biology or Evolution for Human Guidance

Robert Mackintosh

CAMBRIDGE
UNIVERSITY PRESS

CAMBRIDGE UNIVERSITY PRESS

Cambridge, New York, Melbourne, Madrid, Cape Town, Singapore,
São Paolo, Delhi, Dubai, Tokyo

Published in the United States of America by Cambridge University Press, New York

www.cambridge.org
Information on this title: www.cambridge.org/9781108004534

This edition first published 1899
This digitally printed version 2009

ISBN 978-1-108-00453-4 Paperback

FROM COMTE TO BENJAMIN KIDD

FROM

COMTE TO BENJAMIN KIDD

THE APPEAL TO BIOLOGY OR EVOLUTION
FOR HUMAN GUIDANCE

BY

ROBERT MACKINTOSH

B.D. (EDIN.), M.A., D.D. (GLASG.)

PROFESSOR AT LANCASHIRE INDEPENDENT COLLEGE; AUTHOR OF
'CHRIST AND THE JEWISH LAW,' ETC.

London

MACMILLAN AND CO., LIMITED

NEW YORK: THE MACMILLAN COMPANY

1899

DEDICATED

TO

The Reverend Principal Scott,

B.A., D.D., LL.B.,

AND TO

RECENT AND PRESENT STUDENTS OF LANCASHIRE

INDEPENDENT COLLEGE

PREFACE

THE historical sketch and criticism here attempted had its proximate origin in two consecutive years' work with a senior class of sociology at Lancashire College. In 1896-97, essays were prescribed on topics suggested by Mr. Benjamin Kidd's *Social Evolution;* while the seniors of 1897-98 attended lectures covering rather more ground. The material thus collected has been again revised and again considerably added to. The literature of the subject is always growing. Some books of consequence, old or new, must have been overlooked. Still, it is hoped that the subject itself has well-defined limits. The appeal to biology, outlined by Comte, newly defined and emphasised by Darwinism, has now been stated in the most extreme form logically possible. Mr. Kidd's book holds that significant position.

In studying the questions raised, the author has found himself, though with certain grave reserves, more and more thrown back upon philosophical principles learned at Glasgow, above twenty years ago, from the present Master of Balliol College.

I wish to express thanks for kind help on different points; to Professor Henry Jones of Glasgow University; to the Rev. A. Halliday Douglas, Cambridge; and, among others, very special thanks to Professor J. Arthur Thomson and Mr. Norman Wyld. Both Mr. Thomson and Mr. Wyld, while busy with important work on the theory of natural selection, found time to give an amateur valuable information bearing on the meaning and merits of Weismann's doctrine of Panmixia.

R. M.

CONTENTS

CHAPTER I

PART I

COMTISM, WITH SOME SCATTERED PARALLELS

CHAPTER II

CHAPTER III

CHAPTER IV

CHAPTER V

CHAPTER VI

PART II

SIMPLE EVOLUTIONISM—SPENCER, STEPHEN

CHAPTER VII

CHAPTER VIII

CHAPTER IX

CHAPTER X

PART III

DARWINISM, OR STRUGGLE FOR EXISTENCE

CHAPTER XI

CHAPTER XII

CHAPTER XIII

CHAPTER XIV

CHAPTER XV

CHAPTER XVI

CHAPTER XVII

PART IV

HYPER-DARWINISM—WEISMANN, KIDD

CHAPTER XVIII

CHAPTER XIX

CHAPTER XX

ANALYTICAL CONTENTS

CHAPTER I

INTRODUCTORY

Science offers to supersede religion as guide to conduct—In form of *theoretical* sociology—Appealing to biology and evolution—Sociology distinguished from politics—From economics—From social philosophy—Akin to evolutionary ethics—Our point of view ; morality taken for granted.

PART I

COMTISM, WITH SOME SCATTERED PARALLELS

CHAPTER II

COMTE'S LIFE AND THE PRINCIPLES OF HIS TEACHING

Comte as founder—His life—His books—The term "Sociology"—"Statics" (cf. Spencer)—"Dynamics"—Divisions of the *Polity*—Comte's religion—The term "Positive"—Four authorities superseded—Comte on psychology—And on ethics—Law of the three stages—Criticism—Transition to the study of Comte's relation to science—He repudiates dogmatic atheism and materialism—His scale of values in the hierarchy of the sciences—Spencer's criticism.

PART II

SIMPLE EVOLUTIONISM—SPENCER, STEPHEN

CHAPTER VII

DARWINIAN AND SPENCERIAN CONCEPTIONS OF EVOLUTION— DARWIN

Evolution came as a surprise—Darwin deals with biology—With species only—Taking " Struggle " from Malthus, he perceives in it (Natural) " *Selection* "—A true cause, but minute ; an immensely slow process —Compare the replies to Malthus—*Sexual Selection* accelerating— Or *Use-Inheritance*—But too much Lamarck, making variation not " casual," but purposeful, would render unnecessary the " Selective " action of " Nature "—Recent doubts as to use-inheritance.

CHAPTER VIII

DARWINIAN AND SPENCERIAN CONCEPTIONS OF EVOLUTION— SPENCER

A cosmic philosophy—Resting on correlation of forces—And on hypothesis of organic evolution—Emphasising natural (physical, material) law— Darwinism as a cosmic philosophy ? Alexander—Cf. Lotze—Cf. Fiske —Spencer values true use-inheritance as accounting for *a priori* know-ledge—But natural selection is *not* the source of his *laissez faire* doctrine ; he looks forward to a future " balance "—His relation to embryology—*Evolution* means growing complexity—In terms of matter—Two other phases—*Dissolution* as death—As catastrophe— *Equilibrium* is theoretical and prophetic—Spencer's sequence of the three phases—Criticisms : on the assumed *beginning* of the process— On its *isolation*—On *equilibrium*, as involving a different point of view—Reason is more than a new phase of complexity—The whole process breaks up into a series of separate evolutions in complexity.

CHAPTER IX

MR. SPENCER'S THREE DOCTRINES OF HUMAN WELFARE

Goodness is *more evolved* conduct, *i.e.* is " wisdom "—An appeal to (cosmic) history !—It is *balance*, of egoism and altruism—An appeal to economics and to (hedonistic) psychology—It is *individual freedom* —An appeal to rights, and to (human) history, emerging from militarism—For which Spencer feels an exaggerated dread—Spencer masses facts rather than unifies knowledge—The "social organism" is only a phrase with him.

CHAPTER X

MR. LESLIE STEPHEN'S "SCIENCE OF ETHICS"

Stephen a utilitarian—Who came to believe in evolution as a scientific fact—Begins here with facts; ethical judgments exist—Organisms seek maximum efficiency—If social "tissue" is "organic"—Then ethical laws may be the conditions of maximum social efficiency— (Nature cares for individuals)—Nature says, "Be strong!"—Ethics says, "Society, be strong!"—The ethical is the *typical* society, and *therefore* ethical judgments are binding—But the type is actual, not ideal!—Society is a complex whole, changing while its parts are unchanged—Criticism—Sanction for individual goodness lies in sympathy merely—Sometimes we are too good for our own interests! Compared with Comte, lacks *authority*—With Spencer; calls "health" the ideal, and ridicules "balance"—With Darwin; not struggle of individual with individual, but of individual with society—With Utilitarianism; discourages the calculation of consequences—Most of his positions may be accepted in a deeper sense.

PART III

DARWINISM, OR STRUGGLE FOR EXISTENCE

CHAPTER XI

"DARWINISM IN MORALS"—MISS COBBE'S PROTEST

Darwinism may be applied to morals by analogy—Or, as here, by explaining man's evolutionary origin—Miss Cobbe attacks Darwin's explanation of the rise of morals out of intelligence *plus* sympathy— And the hypothetical palliation of murder—Little trace of *natural selection* in Darwin's ethical statement—Darwin's analysis may be accepted, not his view of reason.

CHAPTER XII

DARWINISM IN POLITICS—BAGEHOT

Applies Darwinism by analogy—Evolution *transforms* imperceptibly—
By nerve tissue in our case; but nothing depends on this assertion
of use-inheritance by Bagehot; it is a mere illustration—Not *ethnological*, but *political* questions—Problems both of *progress* and of
differentiation—1st, Custom as the remedy for primitive wildness in
the "fit"—Criticism—2nd, Customs winnowed by the test of war—
3rd, Free discussion—Race-blending, etc., as minor factors—Three
limitations on the Darwinian principle in Bagehot's application
of it.

[Note B. On Professor Ritchie's *Darwinism and Politics*—Inconsistency between the different essays—One interesting hint.]

CHAPTER XIII

DARWINISM IN ETHICS—PROFESSOR ALEXANDER

Fusion of idealism and naturalism—Moral judgments are facts, but the
assertion of free-will is absurd—Criticism; capricious; ignores the
content of moral judgments and the germ of a system in them—
Punishment grouped with dynamics?—*Statics* are truly, though
imperfectly, moral—Goodness is a twofold "equilibrium"—This
doctrine is enforced against other definitions—In the *Dynamics*
equilibrium is revealed as endlessly changing, and is called "compromise"—Ideals compete like organisms for survival—Criticism;
not (*a*) true Darwinian struggle, nor (*b*) true extinction—The new
ideals are not wholly new—Ideals are complementary—So far as he
Darwinises he is false to morality.

CHAPTER XIV

REACTION FROM DARWINISM—HUXLEY

Reaction as to ethics—Due to the vision of *struggle* and *pain*—Not
sympathy, but justice is essential—It must suspend *outright* the
cosmic process—Older evolutionism (Greece, India) gave no guidance
—Criticism; nature and spirit are opposed—Yet connected, and
reason fulfils the cosmic process by transforming it.

CHAPTER XV

REACTION FROM DARWINISM—DRUMMOND'S "ASCENT OF MAN"

His precursors—His sympathy for Spencer—His Comtist terminology—
Seeks a *biological basis* for altruism—Corrects Darwin—Not like
Miss Cobbe—Largely like Huxley—But seeks a fairer statement of
the facts—Brings in a second biological function (out of three !), viz.
reproduction—Wallace on the selection of reason—Leads up to
doctrine of "*Arrest of the Body*"—Cf. Cleland on the human skull
—Emphasis on *maternity* and weakness of human infant—Criticism ;
"egoism" and its struggle purely evil ?—Or male sex with its
justice ?—Is domesticity = sociality ?—Has Drummond shown a
factor in progress ?—A better philosophy claims *all* nature for God.

CHAPTER XVI

REITERATION OF DARWINISM : ELIMINATION MADE ABSOLUTE— MR. A. SUTHERLAND

A strong book with some weaknesses—Works out the origin of moral feeling by natural selection—Restates Drummond-like position as
Darwinian (?)—And exemplifies "arrival" of forms—*Biology ;* fitness
to survive—*And* to breed *and* rear—Quantity first relied on—Then
quality—This develops sympathy—Which becomes serviceable—
Anthropology ; everything depends on the approaches to monogamy—
Sociology ; progress is by elimination of the inferior—Even when it
seems to find more rapid means—(Yet he allows *some* progress by
imitation !)—*History ;* retrogression is possible !—For he hates all
militarism—On the whole he does not believe in history—Or in
reason—*Ethics ;* Has dealt only with one-half of goodness !—Egoism
must balance sympathy !—Balance will grow automatic !—Criticism ;
no right to call sympathy *moral*, if only half of morality—Nature
does not select one quality at a time !—Selection said to have worked
—Not *true* natural selection though—Why is goodness not automatic already ?—Do beauty and goodness exist, or do they not ?—
"Yes and no !"

CHAPTER XVII

THE METAPHYSICS OF NATURAL SELECTION

I. Chance in relation to purpose, as *accident*—As *absence of design*—In relation to law ; as *blind* law—As *blind combination of laws*—Compare with the last the scientific or mechanical view of the world ; a number of separate substances ruled by a number of independent laws—Good enough for science, not for philosophy—Darwin ought not to assume things as *really* disconnected, merely because he has not *needed to investigate* their connection—As if organism and environment were accidentally brought together—Or as if organism and organism were *mere* rivals—(They *are* rivals !)—Or as if force and force were disconnected ?

II. Darwin treats *variation* as casual, *i.e.* as a thing with no bearing in itself on the purpose of the species—His theory *allows* this assumption—But does not *prove* it—We all habitually understand the theory in that sense, *e.g.* in contrasting natural selection with use-inheritance—On the fact, evidence is wanted—Conceivably variation may choose very irregularly between many fixed possibilities—This seems to point back to disconnected laws, as in last section.

III. Even on Darwin's own view he is hardly entitled to call the process of evolution natural selection—Aggregate range of possible variation is fixed by the nature of the material—Two agencies must be taken together—Of the two the varying organism, not the blindly selecting environment, seems the better to account for rise of new qualities—Summary of I. II. III.

IV. Kinds of natural selection, A, B, and C—B exists !—If organic evolution is a fact, C exists !—Accelerating any other evolutionary force that may exist, and of course involving B—If A is found *alongside* of C, A must have a separate field where C cannot enter, else inconsiderable—Natural selection (C) lasts as long as nature is nature —Even along with (the more rapid force of) animal intelligence— True reason checks it—Does natural selection ever work by itself (A) ?—Higher animals with fewer births evolve as quickly as lower; has a new force arisen ? or was natural selection never the leading force ?—[Can we regard *intelligence* as the new evolving force ? Dr. Mellone assumes its operation everywhere !]

V. Can natural selection apply to men ?—*Biologically*—Struggle with beasts is over—Famine (A) is rare, and of doubtful tendency— Pestilence (C) does harm—Vice (B)—Crime (B)—War (selects the

wrong way)—Religious celibacy (*ib.*)—Summary—*Sociologically*—
Mr. Kidd's insistence on struggle is really biological; is unproved;
is not an insistence on natural selection—*Ethically*—Mr. Alexander's
competition of "*Ideals*" is exaggerated—And itself implies reason
and sympathy—Mr. Sutherland's elimination of evil *doers* ignores
positive causes of moral progress—Exemplified typically in Jesus
Christ.

VI. If natural selection does not operate where reason and conscience
exist, it yet may *originate* them in the loose and incorrect sense in
which natural selection is said to originate things!—If reason, etc.,
were, as most suppose, evolved and selected—How selected?—Have
adjacent races died out?

VII. Other idealist views—Professor Ritchie praises natural selection
more fully, in vague terms and in some passages—Mr. Sandeman
rejects it, because he believes in the teleological perfection of every
organism—But is it possible to get over the impression produced by
rudimentary organs?—It is enough if the *whole* of nature is *good*,
and its *parts relatively fit*—Dr. Stirling believes the casual variation
which makes an individual can never make a type—Is it certain
that every individual is born differentiated?—Or that *any* differences
are incapable of growing by cumulation into a type?—Possible value
of the hypothesis of natural selection, even if a fiction.

PART IV

HYPER-DARWINISM—WEISMANN, KIDD

CHAPTER XVIII

A "FAIRY TALE OF SCIENCE"?

An intenser assertion of struggle—Not on ground of experiment; evidence
is ambiguous—On ground of a theory of heredity—Darwin's theory
(Pangenesis) assumed derivation of embryonic qualities from qualities
and tissues of parental organism—Use-inheritance possible or probable on this view—But "Atavism" forced the concession, some
"gemmules" had passed on undeveloped from earlier generations till
they found their chance—Galton's figures for resemblance to ancestors
—Hence theories asserting "continuity of the germ plasm"—Parable
of the hierarchy—Galton ("Stirp") does not absolutely deny the

possibility of use-inheritance—But in Weismann's earlier and more consistent views, founded on by Mr. Kidd, amphimixis is the only cause of variation—Extrusion of one of the " polar bodies " securing (?) non-identity of all offspring of same pair—Permutations and combinations of qualities of unicellular organisms—Nature selecting fittest adults, and in them best germ plasm—Parable of the suckers—Of the Nile—No new quality arises, but amount of each telling quality increases—Qualities arose originally, Lamarck fashion, from environment, when unicellular life lay open to its pressure—Unicellular organisms (propagating by fission) and germ plasm are potentially immortal—Correlation alleged between sex and (natural) death ; now sex is absent from the unicellular world—Natural selection might account for the *predominance* (if not *origin*) of sex if Weismann would assume the necessary competition—Romanes alleges that natural selection might account for predominance of habit of dying natural death ; but would not *death by violence* sufficiently prevent any race (immersed in the struggle) from falling into wholesale decrepitude ? —*Origin* of sex and death a mystery ; or " chance " variation ! or effect of molecular constitution of germ plasm !—Weismann's appeal to " natural selection," while he denies " struggle," is metaphysical in the worst sense—Recapitulation, and note of some of Weismann's changes of opinion before 1893—Especially this change : ENVIRONMENT MAY DO SOMETHING TO MODIFY GERM PLASM !—Making true use - inheritance conceivable, though not inevitable — Mr. Kidd is anachronous—Panmixia, the absence of natural selection, is held to involve indefinite retrogression ; important ; questionable.

CHAPTER XIX

HYPER-DARWINISM IN SOCIOLOGY : STRUGGLE MADE ABSOLUTE— MR. KIDD

Resemblance to Comte—Intenser emphasis on biology [cf. Mr. Platt-Ball]—(1) Panmixia = degeneration is inconsistent with dreams of socialism or of final balance—Selfishness, however, may not care for remote consequences—[Ought Panmixia further to imply extinction ?] Also, social " statics " are blotted out—And evolution becomes almost identical with progress—Could not Mr. Kidd save many essential positions without this assumption ?—(2) Next, if progress implies struggle—And selfish reason makes unwilling to struggle for good of the race, supernatural counterpoise of religion must help, as hitherto—

Now, Weismann had riddled his own position with qualifications—Kidd also appeals to biology by a doctrine of the social organism ; but everything here depends on philosophy, not biology—(3) First, the doctrine of reason ; reason is formal, as with A. J. Balfour, Darwin, Drummond—For Mr. Kidd also holds that biological law applies without a break to rational man—Yet reason *disturbs* process of evolution—And Bagehot, Stephen, Drummond have noted other changes due to it—Can it be wholly evil ?—Balfour and Kidd repudiate Kant or Coleridge's deeper sense of " reason "—But they cannot avoid such sense if it lies in the word and in the fact—(4) Secondly, doctrine of religion as anti-rational—Not = " future judgment" ; that is rational !—Can we believe the irrational ?—Does not Kidd tamper with Christian equalitarianism ?—Biologically ; variation *may* be purposeful and professive—Historically ; reason *is* progressive ; by rational methods—Religion its fulfilment—It needs a force to give it motive and constancy.

CHAPTER XX

SUMMARY AND CONCLUSIONS

Self-contradictions—Comte is arbitrary—Biology has been reinforced by evolutionary theories, yielding different forms of sociological doctrine —1. Analogy, without struggle ; Stephen—2. Continuity, without struggle ; Spencer, Alexander (partly)—3. Analogy of Darwinism ; Bagehot, Alexander, Ritchie (?)—4. Continuity of natural selection ; Sutherland, Drummond (?), Kidd—None of these wholly succeed ; old authorities will return !—Or idealism, which is compatible with the old authorities, may give us a more satisfying doctrine of evolution—What have we been taught ?—(1) A social organism exists—Idealism reinforces this lesson—(2) Struggle *has* been useful ; *will* it not be ? as discussion ? as competition ?—In light of idealism this seems possible—Of fact, probable—Must not exaggerate its place ; it is subordinate in life of reason—[Mallock]—Finally, does *progressiveness* of evolution make it a guide to conduct ?—Difficulties in *biology ;* environment constant ?—Some forms have stopped !—Some never started !—Differentiation plainer here than progress—Reason makes for progress in *history*—Is it all-sufficient ? (Mill, Buckle)—Ancient civilisation failed—Morality and Christianity must safeguard modern civilisation.

CHAPTER I

WHEN the French garrison left Rome in 1870, fears were openly expressed that anarchy would break out, but the Italian troops were promptly marched in, and all went quietly. Religion is supposed to be a retreating force in modern life, and many, even of those who are no friends to religion, suffer grave apprehensions as they look forward to a state of society emancipated from all religious restraint; but others tell us that science will find a remedy. Religion may go off duty, but science will take its place. Never was this conception more confidently advanced, or with more elaboration, than in the first founding of sociology under its present name.

We must clear the ground, however, by a distinction. It is theoretical sociology that we have in view,—a coherent, deliberate body of doctrines, making, among other claims, the startling claim which we have noted above. Much that goes under the name of sociology is matter of quite a different kind. We may call it practical sociology, and we may describe it as a some-

B

what formless mass of good intentions. In detail it offers many valuable suggestions ; scientifically it is a thing of naught. If we were foolish enough to busy ourselves with it in this discussion we should be embarking on unknown waters, possibly upon a shoreless sea. We shall therefore take nothing to do with practical sociology. It is the science or alleged science of sociology that claims our attention.

One outstanding feature of this science is its connection with biology. In the early days of modern history, mathematics stood out in sharp and isolated relief as a well-finished and well-formulated science. Hence an impression got abroad that other sciences were to be perfected by treatment on mathematical lines. Spinoza's *Ethics*, with its array of definitions, postulates, and axioms, and with its pedantic series of syllogisms, is only the most celebrated and most notable among many similar attempts. In our time, biology seems to have cast a like spell upon the minds of not a few. It is biology nowadays which threatens to invade and annex every province of thought. Already in Auguste Comte, the founder or the godfather of sociology, biology counts for a great deal, and subsequent evolutionary speculation has enlarged its claims to infinity. If we achieve anything in this essay, it will probably be in the way of finding a definition (or a cluster of definitions) for the fascinating term "evolution," and in forming an estimate of the value which it, or which they, may possess as affording guidance to human conduct.

Let us further clear our thoughts before beginning our investigation by endeavouring to "place" sociology, provisionally, in relation to other kindred sciences.

In contrast with *Politics*, sociology deals with the

informal or unintended[1] results of human association. In ancient days the line of division scarcely existed. The conception of a *natural growth* had never been applied to society. Speculation in early times was exceedingly sanguine, and counted upon refashioning society at its pleasure. We have learned from age-long experience that human nature is not so easily tamed or managed, even by those who try to manage it for its own good. We turn away incredulously from stories of a lawgiver who stamped his own personality and ideas upon many generations. Perhaps we go too far in our recoil from the ancient belief in the powers of the wise man. He may not always have been a myth ; his results might even be repeated. And yet, essentially, we are in the right. " All the world," as we say, is wiser than anybody in the world. To take a more definite example, the House of Commons is alleged to possess better taste than any one of its members. Our modern attitude is partly fatalism, but it is partly religious faith.

A second science may be thought of, which deals with the objective and involuntary tendencies of social conduct—economics or political economy. This was on the ground before modern sociology, and Comte, who gave the latter science its name, and claimed to be its author, regarded economics as a fragment of social science, wrongly studied in isolation from the rest, and therefore resulting in mistaken practical conclusions. In point of fact, one of the great difficulties or ambiguities of sociology arises no less plainly in economics. How make the transition from study of

---

[1] Compare Mr. Mallock's definition of evolution as " the reasonable sequence of the unintended " (*Aristocracy and Evolution*, p. 97), quoted in our closing chapter.

facts to maxims for conduct? In other words, is political economy an art or a science? The accepted view nowadays regards political economy as a science —the science of wealth; and in spite of Comte's protest, it is recognised as a distinct science, independent, in a sense, of sociology; and that, mainly because more definite conclusions are possible in regard to wealth than in regard to the wider social interests of mankind. On the other hand, it is fully recognised that, if you wish to frame maxims for conduct, you have to take much into account besides the economic tendencies of action. And it is also confessed that in its "palmy days" political economy had identified itself with a system of individualism — with a hard doctrine of individual rights, more especially rights of property— which may well be thought a menace to the public interests. Nevertheless—such is the irony of circumstances! — practically the same system has reappeared in all its stringency in the form of Mr. Herbert Spencer's sociology.

Thirdly: Professor Mackenzie's *Introduction to Social Philosophy* adds another distinction—that of social philosophy in contrast with social science. Sociology claims to rank as a science; Mr. Mackenzie, who is entitled to respect, both on his own account and as representing generally the position of the great Hegelian or Idealist school, conceives that there are philosophical positions presupposed in social science which need separate discussion. In consequence or partly in consequence of this, Mr. Mackenzie's book does not aim at giving us a body of social doctrines, but at vindicating on philosophical grounds what he regards as wholesome social principles. The main significance of this, we think, is as follows, that, in contrast with the school which seeks to reduce social well-being to a problem in science, in analogy as far as may be to physical science and in close connection with it, there is another school, not less attached to a doctrine of corporate well or ill, which finds the highest authority in regard to human conduct in metaphysics.

Fourthly: we might speak of the relation of sociology to ethics. But here the floods threaten to break loose and drown us. Here we come face to face with the question already mentioned—the question of the transition from science to art; from noting how things happen to declaring how they ought to happen. Without enlarging further upon that topic at this stage in our discussion, we may at least call attention to the fact that historically there has been a very close kinship between sociology and ethics. Their problem is almost, if not altogether, the same; the answer formulated is sometimes labelled "sociology," at other times "ethics," as on shipboard the jam is sometimes described as raspberry, sometimes as plum, sometimes, it may be, as guava, yet in all you taste the monotonous flavour of apple, or of burnt sugar. Not less alike to each other are evolutionary ethics and evolutionary sociology. Thus—to anticipate for a moment—sociology was originally formulated by Comte as the true guide to conduct, the new authority, destined to supersede both ethics and religion.—He modified this position in later days, as we shall see, but only within limits, and at the outset it was announced as we have given it.—Sociology offered to guide man with the help of biology; society was an organism; man was a member in the organism; a part, not the whole; essentially dependent on the whole, and bound to serve its interests. This conception reappears in Mr. Spencer; he works out its suggestions in his own way, which is not Comte's; but still he appeals to the analogy between society and an organism; and he calls the discussion sociology. But when we turn to Mr. Leslie Stephen's *Science of Ethics*, we meet with identically the same discussion. True, Mr. Stephen prefers the expression " social tissue " to the expression " social organism," but the difference is essentially one of detail, and does not affect the question before us. We are still working the biological analogy, yet, if you please, this is ethics we are working at. The brand, no doubt, is different; the liquor is the same. Spencer has elsewhere and in different form his discussion of ethics; Stephen's ethics run parallel, not to Spencer's ethics, but to Spencer's sociology. Again, Professor Alexander's *Moral Order and Progress* is, as the name implies, an ethical discussion, yet the author finds it impossible to discuss the problems of personal ethics apart from the relation of the individual to society, and his book is penetrated throughout with biological and evolutionary suggestions, most of all with the Darwinian struggle for existence. But such suggestions meet us at every hand in modern sociological discussions; nay more, such suggestions it was the professed business of sociology to supplement and apply to human life. It is plain, therefore, that sociology and ethics, as sociologists generally conceive of sociology and of ethics, cannot be separated from each other. Some forms of ethical thought will wander far from the line of treatment proper to us in this essay. But, wherever you have these two things—an interpretation of *duty* as the debt which man, the individual, owes to society; and secondly the appeal to phenomenal fact as the only safe or real authority—there sociology and ethics must necessarily approach, intertwine, or even coalesce. And therefore it would mutilate a study of sociological theories, not to include in our review those ethical systems which are plainly of the same house and lineage.

Every argument proceeds upon certain assumptions; and it may be as well to confess at the outset what is to be assumed in the following essay, viz. the trustworthiness of the moral consciousness, or the reality of the distinction between right and wrong. This test will not be formally set aside, except by a few wild thinkers; but it may be objected that assumptions ought to be vindicated, ought to be justified. Very true; our test needs justification by philosophy, and we believe that philosophy can do the necessary work, but not here. We cannot incorporate *en passant* a body of metaphysical prolegomena to ethics. We must be allowed to let our point of view stand as an assumption.

Looking at matters thus, although we seek to learn from the theories reviewed, and especially from the interesting and valuable details which they have collected, yet our analysis will necessarily to a large extent be hostile.

First, we ask whether the various theories agree with each other? And on this Mr. Benjamin Kidd, himself a sociologist, tells us that the sociologists are hopelessly divided in their attempts to furnish practical guidance. The science was to have been founded by Comte fifty years ago and more; Mr. Kidd seems to think it still needs founding by a new recurrence to biology. It is plain, therefore, that the appeal to fact has not yet done for the study of society what it promised to do. Neither theologians nor metaphysicians could have been more hopelessly at issue among themselves than the votaries of fact have been and still are. Secondly, we ask

whether each author is so much as self-consistent?
Thirdly, we ask, granted that we learn some fresh truth,
is it taught us authoritatively by science, whether by
the science of biology or by some other? or has natural
science merely suggested parables to the moral judgment? These formal or logical tests pretty well clear
the ground. A remainder of our theories, however, is
overthrown (fourthly) by the final test, by the touch-
stone of the moral consciousness.

Positively our argument can hardly be said to go
beyond this point, that if biological clues are to afford
guidance for human conduct, they must be supplemented
by clearer moral and religious light, and in philosophy
by some scheme of metaphysical evolutionism, marking
a transition perhaps from " Darwin " to " Hegel."

PART I

COMTISM, WITH SOME SCATTERED PARALLELS

CHAPTER II

COMTE'S LIFE AND THE PRINCIPLES OF HIS TEACHING

Comte as founder—His life—His books—The term "Sociology"—"Statics" (cf. Spencer)—"Dynamics"—Divisions of the *Polity*—Comte's religion—The term "Positive"—Four authorities superseded—Comte on psychology —And on ethics—Law of the three stages—Criticism—Transition to the study of Comte's relation to science—He repudiates dogmatic atheism and materialism—His scale of values in the hierarchy of the sciences—Spencer's criticism.

ALONE perhaps of all sociologists, Comte may claim to have his life studied, however briefly, as an integral part of the gospel he teaches.

Auguste Comte was born at Montpellier in 1798. He was early distinguished for his mathematical ability; also for a refractoriness to authority, which led to his expulsion from the Polytechnic School of Paris. In 1818 he met St. Simon the socialist, and became for six years his close friend and disciple; but the alliance was broken off by a violent quarrel, never to be healed. In 1825 he married. The union proved conspicuously unhappy, and ended in a separation in 1842. In 1826 he began lectures upon his system of philosophy; and though they were interrupted for a time by an attack of insanity, the lectures attracted great attention. Between 1830 and 1842 they were published in six volumes under the title of *System of Positive Philosophy*. While working for fame or usefulness by developing his system, Comte worked for bread and butter by the exercise of his mathematical talent, mainly in the service of that Polytechnic School from which he had been expelled in his student days. His eminence as a heresiarch cost him his connection with the school; and thereafter he lived by his earnings as a private tutor, or by the gifts of his devoted disciples. In 1845 he became acquainted with his Egeria, a lady named Clothilde de Vaux, with whom he fell passionately in love, and to whom he looked back with passionate regret till his death in 1857, the lady having lived only one year after making acquaintance with Comte. There was no stain on their friendship, though it was the occasion of a good deal of folly upon Comte's part. In his later years, 1851-54, Comte published the second part or second form of his system, the *Positive Polity*.

We do not attempt to mention other works, but it is necessary to say something about the *Philosophy* and the *Polity*. The earlier treatise, the *Philosophy*, was an encyclopedia of scientific knowledge, as it then existed, crowned with the first rough sketch of the science of sociology. It was condensed in an English translation by Harriet Martineau, a translation which was afterwards retranslated into French, as being an improvement upon Comte's own statement. This may be called our English tit-for-tat in exchange for Dumont's relation to Bentham. The book was recently republished in English, when an able reviewer[1] protested against the absurdity of offering the reading public the

---

[1] In the *Manchester Guardian*.

science of fifty or sixty years ago. The *Positive Polity*,
on the other hand, is sociology from beginning to end ;
looking back, as we shall see, to the survey of the
inferior sciences made in the *Positive Philosophy*, but
working out its own problems on the grand scale.

In the earlier book we have the two main divisions
of sociology—first, social statics, or the conditions of
social order ; these are treated briefly ; secondly, social
dynamics, or the historical laws of social progress in the
past.

All three names are somewhat singular. The name
sociology—Comte's own coinage—is a hybrid term,
partly Latin and partly Greek. Social statics, again, is
used in a different sense from that of Mr. Herbert
Spencer's early treatise. With Spencer, social statics
refers to a future Utopian period, when egoism and
altruism are perfectly balanced ; a millennial age, when
"that great disturbance of human nature, which the
churches call sin," has been left behind. It therefore
corresponds to the "absolute ethics" or "ethics for the
straight man" of Mr. Spencer's later system—a fresh
proof, if further proof were needed, that ethics and
sociology are only diverse names for the same product,
as production is carried on in the schools of empirical
sociology and evolutionary ethics. In the light of
science it would seem that Comte's use of the phrase is
much better justified than Spencer's. Mechanical statics
discuss the conditions of stability in actual life, not in
some ideal world, where the properties of things have
been modified out of all recognition. Lastly, the phrase
social dynamics ought in accuracy to be social kinetics.
By rights the name dynamics covers the whole field of
mechanics, studying the conditions both of stability and
of movement, and thus including as its two branches

statics and kinetics. As for the name mechanics, it is
usually extruded by men of science from the field of
theory, and confined to practice. However, the words
dynamics and dynamical are so identified in sociological
usage with that half of the subject which deals with
motion, or, in other words, with historical change and
growth, that it does not seem wise to attempt to disturb
the inaccurate but well - established phraseology of
tradition.

The later book, the *Polity*, not only has a fuller
discussion of sociology, but a greater number of topics
or heads or subdivisions. First, there is a general
sketch of Positivism. Secondly, there is an outline of
the principles to be fully developed in what follows.
Thirdly, there is an account of Social Statics, *i.e.* of
permanent conditions of social order ; very much fuller
than in the *Positive Philosophy*, and therefore not
merely naming or sketching in brief the Family, the
State, and the Church or Humanity, but treating the
last specially at greater length, and adding discussions
upon Language and upon Art. Fourthly, we have
Social Dynamics, Comte's Philosophy of History. This
had been given with disproportionate fulness in the
early treatise ; but its discussion is a good deal enlarged
in the later volume, though other points are still more
enlarged. Lastly, there is the Polity of the future,
dogmatically detailed upon Positivist lines. It is plain
that such a programme affords plenty of scope for
repetition and reiteration. Comte makes full use of his
opportunities. We must remember that Comte had
already in view the composition of the *Polity* when he
issued his *Philosophy*. It is characteristic of the man
to grind his few leading ideas round and round and
round again in his own and in his reader's mind. A

division or a generalisation is never expounded once for
all; we shall meet with it again as a subdivision in a
different section. This is a failing which leans to
virtue's side, but its scale is positively gigantic in
Comte.

Along with the difference in scale, and in precision
of semi-political or legislative detail, there is to be
noted a difference, up to a certain degree, in the
animating spirit. Both treatises rely upon Comte's
hierarchy of the sciences; both rely upon his historical
law of the Three Stages; and both of them are affected
by his belief that the heart ought to rule the head, or
the intellect to be the servant of the affections. But
the last point certainly counts for vastly more in the
Polity than in the *Philosophy*. Between the date of the
two treatises the church of humanity, as represented
by its prophet Comte, had developed a whole system
of worship. Some have regarded the two stages of
Comte's thinking as flatly contradictory of each other.
It seems better to recognise that, at every stage, there
were diverse currents of thought or "streams of
tendency" mingling in Comte; that he was perhaps
divided against himself, habitually inconsistent, continuously self-contradictory. Certainly it is hard to
reconcile the view that the heart is to be master of
the intellect, and its result, the sentimental worship
of Humanity, with the appeal to mere phenomenal
fact. Yet Comte and the Comtist elect are conscious
of no self-contradiction. Both demands are merged in
the blessed and magical word—Positive.

What is it to be Positive? In French, the word
may have a special history, giving it a richer connota-
tion. In English it has no such distinctive position;
it is merely the opposite of negative, or sometimes of

natural, as when we contrast positive law with the obligations of natural law. Perhaps a combination of these two senses may suggest the Comtist view, especially if we can light up the result with an unspeakable glamour of love and complacency. Comte prefers positive historical institutions to what he regards as metaphysical dreams of natural law or natural rights. He prefers real facts to fictitious or ideal fancies. Yet the fictions had their use. They helped to clear away the mediæval system, in doctrine and polity, when it had grown obsolete. More than that the spirit of the Revolution—or, as Comte would say, the spirit of the Reformation and of the Revolution —could not possibly accomplish. But more is now demanded. That negative service has been done. We must be positive. Back then to the facts; if we appeal to the right facts, in the right spirit, we shall positively save society ; positively, we shall !

The old authorities, whose defeat Comte usually takes for granted, were at least three or four in number. There was religion ; supernatural religion ; what students call the positive religions of the world, claiming, many or most of them, to come by revelation. These had played their part in promoting human or social well-being during the theological stage of history, but they were long ago effete ; the metaphysical stage had super-seded them, and it in turn was now yielding to the final or positive stage of knowledge. The other three authorities are all metaphysical, and on that ground are disowned by Comte ; metaphysics proper, the intro-spective method in psychology, and intuition. As it happened, these various alleged authorities had presented themselves in alliance to confront the assaults of modern Agnosticism ; and, as Comte believes, they had all been overthrown. The third, the introspective method in psychology, is perhaps not strictly an alleged guide to conduct; but it stands in very close alliance with the fourth. If simple interrogation of consciousness teaches us truth in one great department of knowledge, then simple interrogation of the voice of conscience may well be expected to teach us duty, and guide us safely in action. Comte, a more thorough-going empiricist and phenomenalist than his English colleagues, the Mills and Spencers, is resolved to have nothing to do with the psychology of introspection. Psychology is either a department of physiology, phrenology perhaps; or, as he says in his later treatise, sociology is the true psychology, *i.e.* sociology gives us the one true doctrine of man. On the other hand, it was the earlier treatise which offered us sociology in lieu of ethics,—which, as we may say, carried its aversion to intuitionalism so far as to blot out of being the science which intuitionalism had so deeply infected. The later treatise recognises that a science of morals ought to handle the problems of personal conduct, in the light of the conditions of social well-being established or defined by sociology. As being more complex, the discussion of personal duty in morals—a treatise which Comte never was able to compose—is placed by him later than sociology in his list of the sciences.

Beyond this statement of his alleged Law of the Three Stages, Comte does not argue in favour of his agnostic background. He takes it over from his predecessors in the business of speculation, empiricists and individualists of the ordinary type. Once he refers to Kant, telling us that Kant had had a very fair inkling of the biological view of human knowledge as a thing absolutely relative to its environment—being partly due to the activity of the organism, partly to the reaction of the environment; the two elements mixing in a way that defies us to decompose them, and that forbids us to regard man as capable of possessing absolute truth. But usually Comte is content to let history, as he understands history, tell its own tale. Once, mankind aspired to penetrate to the knowledge of causes. The race devoted itself to a theological interpretation of the world. First came Fetishism; every object in nature, every part of the mighty whole, was held to be alive, just as man himself is alive. Unlike the writers of to-day, who generally identify Fetishism with Animism—in the most approved sense of that slippery and misleading word—Comte has no intention of admitting that primitive mankind believed in spirits, temporarily or permanently connected with the Fetish. Not so; Comte regards a belief in the soul as belonging to a much more sophisticated state of mind than that of the amiable fetish worshippers, the first fathers of the human race. Not until the baleful shadow of metaphysics begins to fall upon human thought do we hear of souls in men, or of spirits in nature. To Comte, psychology is a kind of physiology; psychical life is a property of the human body; and, to the fetishist, psychical life was a property of the objects of nature. Again Comte differs from ordinary usage in extending the term fetish to cover any object in nature which might be worshipped—a river, a mountain, a star, the moon, the sun. By other writers, that highly ambiguous and arbitrary word is usually applied only to things which are or which may become private property. Fetishism, as understood by Comte, was regarded by him as the first form of religion. This, again, was part of the legacy to Comte of the Ency-

clopedists and their fellows. Out of Fetishism, according to Comte, grew Polytheism. The change is mainly
attributed to the action of human reason. It came to
be discovered that things which had been regarded as
animated were really inanimate. But the theological
delusion was not yet shaken off; the human mind was
not yet strong enough to go right on to the scientific
or positivist consciousness. Instead of doing that,
mankind invented a set of imaginary beings, called
gods, lurking behind the phenomena of nature. To
the gods were now attributed those activities which
observation would no longer suffer men to ascribe to
stones or plants or unconscious natural forces. Next,
out of Polytheism grew Monotheism. Here again
reason had been at work; as the unity and harmony
of nature was more and more discovered, it became
more and more difficult—at length it became impossible
—to interpret the world as an effect produced by independent or rival agencies. There must be one great
first cause; one great manlike Being. Monotheism
had begun; the last term in the theological development. But the development was to continue beyond
Monotheism, and already, unnoticed, under the domin-
ance of the theological stage, the germs of the
metaphysical stage of mind were developing. Metaphysics, according to Comte, sees through the absurdity
of belief in gods or in God; reason is still active, and
is very strongly impressed at this stage (says Comte)
with the moral difficulties of Theism; but, according
to metaphysicians, all we have to do is to substitute
abstractions for the discredited deities. In the metaphysical stage of thought we take these abstractions
seriously, as if they could give a real and satisfying
explanation of things; but they are only ghosts of

causes, ghosts of gods, ghosts of the real living body
under the style and title of souls,—and so forth and so
forth.   Drugs produce sleep because they have a soporific
virtue.   Life is due to some mysterious intangible vital
energy.   Chief of all the abstractions is Nature.   Substitute *Nature* for the monotheistic *God* and the feat is
accomplished; the transition is made; the first stage of
thought has given place to the second.   With the conception of nature grows up a crop of wild beliefs in
natural laws—he means the jurist's natural law, not the
physicist's—and in natural rights.   These beliefs are
powerful to destroy, powerless to create.   But that is
their use,—to clear away the rubbish and debris of an
obsolete intellectual and social order.   Hence the reign
of metaphysics must be incomparably shorter than the
prevalence of the theological spirit.   Already the new,
the true, the final stage of thought was unfolding itself
in a few rarely gifted minds.   The one solid result of
metaphysical inquiries consisted in the fragments of
science accidentally discovered, either in antiquity or
in the Middle Ages, by minds too finely touched for the
metaphysical dreams which chiefly occupied them.   In
a sense, therefore, science antedated metaphysics.   But
more still, there must have been a leaven of positivism
—*i.e.* of science—even in the earliest fetishist days, if
human life was to be maintained on earth.   And so we
do not wonder to find that society was being built up,
piece by piece, long before sociology was possible.   In
the days of fetishism the family was developed,—the
most essential of all social formations.   Polytheism,
which ushered in the epoch of militarism, witnessed the
construction of the State.   At first, however, spiritual
power and secular power were closely combined.   Either
the State was a Theocracy, in which the priests ruled;

or in subsequent days the military classes, who had
assumed command of the State, kept the priests under
control.   Both of these systems yielded very imperfect
types of the State; yet humanity owed much to them.
The practical wisdom of the priests, and, still more, the
sagacious instincts of secular statesmanship, did a great
deal to counteract the anti-social tendencies of a developed theology.   Instead of dreaming away their
lives in religious joys, or in thoughts of another world
—as their creeds may have demanded—men were dis-
ciplined by their wise rulers to think of the interests
of their country, and to aim at the public weal.   Under
Monotheism once more—*i.e.* under Christianity, or, as
Comte calls it, " Catholicism "—a very great advance was
made through the mediæval separation of the spiritual
and temporal powers.   The empire of the German
Cæsars and the ecclesiastical Papacy stood over against
each other in seemingly hostile array as competitors for
the supreme place.   Really, says Comte, the separation
of theory from practice—for that is what it means from
his point of view—was a decisive gain for human well-
being.   During the same epoch chivalry or defensive
warfare formed a transition stage from the old aggres-
sive militarism to the modern Industrialism.   So much
had already been wrought by the spirit of positivism,
even before it had come to self-consciousness.   But now
science is fully accredited and well grown ; and industrialism, the definitive social order, which corresponds
to science or positivism, the definitive stage of thought,
lies all around us, albeit still in sad confusion.   The long
regency of God is at an end.   The minority of Humanity
has ceased.   We are done with dreams of knowing the
causes of things ; we are content henceforth to register
sequences, and to calculate phenomena, for the practical

ends of human welfare.   Comte has appeared, and, by attending to his teaching, mankind now at last may enter the land of promise.

Of course the value of this historical sketch of the progress of the human mind depends upon the degree in which it is true, and in which its truth can be demonstrated.   It is hardly necessary to say that while it reveals wide knowledge and great power of generalisation, it also contains many assumptions, and much prejudice, and not a little which is now proved to be false. The early history of human religions and human institutions is still indeed extremely obscure.   Many theories are put forward ; none can claim a complete victory. And yet it is not too much to say that Comte's neat little sketch of Fetishism, and its uses, and its successors, must be laid aside among the things which are curious but not serviceable.   However, the question specially before us at this moment is whether Comte's historical survey justifies his agnostic creed.   In support of Comte there is one striking fact to be noticed.   The field assigned to natural law has constantly tended to expand ; supernatural agency, even by those who believe in it, has been put farther and farther back, farther and farther off.   So much Comte may certainly claim to have made good.   But it is still matter for argument whether this really points to the cessation of theological and metaphysical belief.   The question is a metaphysical one, to be fought out on metaphysical grounds.   In his dislike and contempt for metaphysics, Comte offers us merely what one may call historical statistics of the dwindling of faith.   But that is to postpone the question indefinitely.   Till faith in God has died out like faith in witchcraft, history cannot claim to pronounce upon it a sentence of worthlessness.

Or we may propose another issue. Let us consider Comte's appeal to science. If that works out so clearly and satisfactorily as to carry us unhesitatingly with it, then we may feel that Comte has justified his cavalier attitude towards those mighty allies, faith and reason. On the other hand, if Comte's positive construction fails to commend itself, we shall be justified in "considering yet again" the old-fashioned guides to truth and duty, for which sociology was to be a substitute.

Now, first, we must remark that Comte does not absolutely shut the door against faith. While he regards belief in a God as the second-last outworn raiment of human thought, he declines with some indignation to be called an atheist. God, say his disciples, may or may not exist; the question lies beyond the competency of human reason to settle. So, too, the doctrine of a soul separate from the body is assigned by Comte to the last outworn phase of thought—the metaphysical. Yet, if you call Comte a materialist, his facile indignation once more overflows. He belongs, therefore, to the agnostic group. He will neither say "yes" nor "no." But he is filled with scorn for those who say "yes," for he is perfectly and dogmatically assured that we have no right to dogmatise. Moreover, his attitude towards the claims of his rivals looks very differently in different sentences or paragraphs. When he denounces the dreams of theories that transgress the limits of human reason, he speaks in the tone of one who possesses real knowledge through the positive sciences. But, when he explains that mankind is abandoning inquiry into causes, it forces itself with a shock upon the reader's mind that the opposite is the case. It is knowledge that we are surrendering. It is reality that we are forsaking. Our predecessors may have failed to attain real knowledge. For argument's sake take it, if you like, that they failed pitiably. Still there is this to be said, they tried; whereas we, the crowning race, are to give up real knowledge, and to content ourselves with registering useful sequences. We have not awakened from a dream, but rather fallen from a dream into a stupor. This also is characteristic of the whole agnostic group. It is easy to write the words "limitation" or "relativity of knowledge"; but it is hard to work out your meaning so that this relativity or this adamantine limit shall not involve the abrogation and annihilation of knowledge. But those who despise metaphysics far too thoroughly to study it, will always be found rejoicing in scraps of metaphysical "creeds outworn."

Next, we observe that, while Comte appeals to phenomenal fact and positive science, he does not place all sciences upon the same level. He has arranged them in a scale—1st, Mathematics (including Arithmetic, Algebra, Geometry, Mechanics); 2nd, Astronomy; 3rd, Physics (with subdivisions—Sound, Light, Heat, Electricity, etc.); 4th, Chemistry; 5th, Biology or Physiology; 6th, Sociology; to which the *Positive Polity* adds, 7th, Ethics. In the *Positive Philosophy* there is a full review of the state of knowledge regarding the various branches of mathematical and physical science at the time when Comte wrote. This order is regarded as the best order, the right order, the order chosen by the φρόνιμος, the wise and well-cultured man, Auguste Comte. It is not simply an order of initial ease and progressive difficulty. It is mainly an order for study—roughly coinciding with the order of discovery—but principally justified by the statement, that each science presupposes the results of its predecessors, while it marks out for itself a new

field of scientific achievement by detecting new uni-
formities. Before Comte, it is urged, there was no
science of society. Comte learned from biology to
regard society as an organism, profoundly related to
its environment. But that did not establish a science
of sociology. Two luminous generalisations did so—
the Law of the Three Stages, and the Hierarchy of the
Sciences. This illustrates to us the intricate arrangement of material characteristic of Comte's redundant
method. The Hierarchy of the Sciences includes
sociology; but again, the hierarchy is revealed to
mankind by sociology; and, once more, the hierarchy
constitutes one half the title-deeds of sociology, justifying its claim to be ranked with the sciences.

It is a somewhat remarkable development of phenomenalism, this arrangement of sciences, not merely in
sequence, but in a rising scale. It recalls to mind the
great Idealist systems of Germany, so like, and so
unlike, Comte's philosophy. One is not surprised to
find Spencer protesting against the ladder of knowledge,
—protesting that the relation between different sciences
is not one of superiority and subordination, but one of
equal reciprocity, each borrowing from each, each lending to the other. Still, if only because, as Carlyle
said, "speech is linear though character is solid,"—still,
it is necessary to take sciences one at a time,—first one,
then another; the synthetic philosophy itself has a beginning, a middle, and an end. And probably Comte's
view has better justification than Spencer's, though
there is a measure of truth in each. It is true that
borrowing and lending go on between different sciences,
backwards and forwards, up and down; but it is also
true—and the truth is of greater importance—that
high branches of science are dependent on the results

of lower and simpler branches. In spite of the pre-
judices of phenomenalism, a scale of values *will* assert
itself as we deal with the different branches of human
knowledge. Of course Comte had his own explanation
of the origin of this scale of values. It is purely
subjective, a matter of human convenience. To take
things in this order suits us, and therefore we rightly do
so ; for intellectual curiosity is always to be kept in
subordination to the claims of the affections. But how
does it happen that human knowledge, upon the whole,
lends obedience to the demands of the moral nature ?
How is it that knowledge comes to us, imperfectly but
really, in the form of a system, where the later parts
imply the previous parts and carry us further on ? In
other words, how comes it that our subjective synthesis
does not distort the knowledge which phenomena afford,
but rather brings out its inner meaning ? Comte is in
a curious half-way position between phenomenalism, to
which one fact is as good as another, and idealism, to
which knowledge is a thing that objectively and really
grades itself. It is a thin disguise of intellectual helplessness when Comte asserts that we have such a grouping
of phenomena in our knowledge, but that the grouping is
due merely to man's capricious regard for the interests of
his own species. " Facts are chiels that winna ding."
They are not so easily manipulated as Comte implies.

Putting the matter in our own way, we may say
that Comte's positive and constructive teaching has
three sources of light and leading, in which it trusts—

(1) The appeal to Biology.
(2) The appeal to History.
(3) The doctrine of Altruism.

We shall say a few words about each in turn.

CHAPTER III

THE APPEAL TO BIOLOGY

The "social organism" in other writers—In Comte—Idealist supplement to the biological appeal—Professor Mackenzie's statement of the idealist view—Intuitionalist criticism of the appeal—Comte uses a biological parable—Consistent phenomenalism means (if not evolutionism) hedonism—Comtism and hedonism two half truths.

[Note A. On Drummond's *Natural Law in the Spiritual World*—"Biological religion," according to Finlayson—Drummond appeals to biogenesis—His religion is Calvinistic, rather, or Gnostic—His noble zeal for continuity in knowledge.]

BIOLOGY comes next below sociology in Comte's scheme of the sciences. As we have seen, it is somewhat difficult to know how far, upon Comte's own principles, this juxtaposition of the two sciences warrants him in expecting the ideas of the lower science to serve as a guiding clue in the construction of the higher. Let it be enough to say that, whether in obedience to his own principles or without warrant from them, Comte has drawn a good deal from the biological analogy. As far back in time as the secession of the Roman Plebs, the parable of the "belly and the members" is alleged to have taught moral lessons to hot-headed or selfish factions. Again, in St. Paul's account of the Church, we are introduced to an organism in which all the members rejoice or suffer together, sympathising fully with one another. It is an extension of the Christian spirit which leads modern thinkers to apply the same image to the State or to civil society. The contrast has been tellingly drawn between St. Paul's appeal as to a well-known fact—" Ye are members one of another" —and the Greek despair of being able to name any authority strong enough to overrule personal selfishness. When modern thinkers call society an organism, they say in effect, not merely to fellow-Christians, but to fellow-citizens or fellow-men, " We are members one of another"; they say it, counting on a response; and they obtain not a little response, thanks to the spread of the Christian spirit and Christian ethic. Moreover, science takes up the keynote in such a phrase as " the physiological division of labour," a phrase which shows us how the lower science is at times indebted for suggestions to a higher—in this instance, physiology, to the economic branch of the science of society—but which also shows us the reality and the scientific serviceableness of the analogy between the two fields of study.

Apparently Comte himself was aware that biology and sociology in some respects formed a class together, contrasting with the lower sciences. In his little book on Comte, Dr. Edward Caird twice over [1] tells us that Comte recognised even in biology, much more in sociology, the necessity of bringing to a focus that *esprit d'ensemble* for which he pleads, and for explaining the parts by their place and function in the whole, not the whole by the co-operation of mutually independent parts. This spirit grew on Comte more and more. " Humanity," he said at last, " is alone real; the individual is an abstraction." In so far as he appealed to biology for encouragement in such teaching, Comte was following biological clues in the new science of sociology.

Now, if this be so, an adherent of the German

––––––––––

[1] 2nd edition, pp. 61, 132.

idealism will welcome Comte's progress, such as it is. He will think it far better to expound human reason—and what he regards as a creation of human reason, human society—in terms of biology rather than in terms of mechanism, or of "matter and motion." Neither interpretation may be adequate, but Comte's will seem to the idealist much nearer the truth than the other. Only the idealist will lament that the scale of the sciences is cut off with a knife at biology. He thinks life a truer, richer, fuller, worthier category than affinity or force, or any purely physical conception; but he believes there is a higher category still, viz. self-conscious reason. He believes that, while the processes of life may do a good deal to throw light upon the processes of reason, the processes of reason throw back even more light upon the allied yet inferior processes studied by physiology. The idealist holds that reason has gone to the making of all things; that it shows a little of itself in the lower sciences, much of itself in the sciences of biology and physiology, but all of itself in self-consciousness—self-consciousness, which is the open secret of the world, and which does not need to be studied at second hand either in biology or in sociology when we can study it in itself, and in its workings everywhere. Good to use biology as a help, says the idealist; but why stop at biology?[1]

It is perhaps the same position in different words when Mr. Mackenzie tells us that *his* doctrine of an organism (as applied to the social organism) is a metaphysical category. The perfect realisation of unity in difference, the whole in all the parts, each for all, and

---

[1] With an interesting and characteristic modification, Professor Baldwin of Princeton affirms that *Psychology* gives us the true clue to the nature of society.

all for each, is only hinted in natural organisms, but is
achieved in the life of reason and of goodness.   Men of
science need not trouble to tell idealists of supposed
errors in the idealist conception of an organism.   Idealist
philosophers go to science for hints, for rough outline
sketches, for parables; it is to reason they apply for
final and authoritative revelations.   Few animal organisms may display any perfect relativity of the whole to
the parts, and of the parts to the whole.   If you cut off
my head I die.   If you cut off my arm, unless you do it
very clumsily, I do not die.   The head therefore seems
to be a necessary and integral element in the organism;
the arm does not.   Or, again, if a lobster loses a claw
he can grow another.   I, alas! may lose a leg or an
arm, and still survive, but I cannot replace the missing
limb.   Is the lobster the truer and worthier organism?
It cannot do without any one part, and if any part goes
amissing, what has been lost is reproduced by the
remainder of the organism.   Or an organism which, so
to speak, was *all heads*, would seem to be a metaphysically perfect or beau-ideal organism, where every
part was vitally necessary, because each part was implied
in all the rest.   The human organism, happily for us,
does not illustrate the metaphysical category in *this*
phase of perfection.   Yet the category is not irrelevant.
In the healing of a wound physiologists recognise something analogous to the mysterious power by which the
lobster grows a fresh claw.   Thus the parable exists in
nature, but the fulfilment is found in reason and in
conscience.   Far more fully than any members in one
of nature's organisms, " we "—human beings, God's
children—" are members one of another."   Our mutual
dependence is absolute; our life, if torn asunder from
each other, is no human life at all.

A different criticism might be stated by one believing
less confidently than idealists do in the completed scale
of the sciences, while attaching more distinctive importance than they attach to the revelations of the moral
consciousness. Such a one would ask, Is this biological
parable anything more than a covert appeal to the
moral consciousness? Is it anything more than a
fantastic way of saying, " You ought," a masked transition from the " So it is " of phenomenalism to the " So
it ought to be" of ethics? Religion, at least in its
historical forms, has been deposed ; Christianity has been
scouted ; intuition has been laughed down ; philosophy
has been told to vanish with the ghosts before the noon-
tide of science. Yes, but how are you going to bring
men under authority when so many authorities have
been sent packing? It is very convenient if you can
assert the claim, the moral claim, of the community in
the parable of body and members ! This may not be a
perfect moral authority, but it is at any rate an authority,
and in the bankruptcy of the moral consciousness any
authority is better than none. Nay, for Comte it is the
very authority he wants, human and governmental.
Yet this doctrine of the social organism is no pronounce-
ment in the name of facts ; it is a moral dictum,
picturesquely stated in terms of popular science. The
community is doubtless part of the moral authority to
which each man owes allegiance. But the parable of
the social organism would not win the wide acceptance
it does if it were not for the authority of conscience
within, and for the training of conscience by the authority
of the Christian spirit during centuries.

We conclude then that the appeal to biology has
done Comte a very great service. After he had cut
away the foundation of morals he has been able to find

a new foundation in the tacit assumption that individual
men are bound to the service of the common weal; and
this assumption is masked, and made to look like the
statement of a scientific fact, by the process of borrowing
a parable from biology.

Of course it may be rejoined that Comte is much
more true to his phenomenalist assumptions, and that
he is merely appealing to fact when he uses the biological parable. Any one, it may be said, can see that
men are dependent upon society, and that selfishness
leads to unhappiness, not to happiness. That, however,
suggests Hedonism, and Hedonism is strange to Comte.
Hedonism represents the earlier and probably the more
consistent working out of a phenomenalist view of human
conduct; but sociology represents a strong reaction from
it, as from other manifestations of individualism. Probably it will be admitted to-day in most quarters that
J. S. Mill failed logically in his generous attempt to
establish the claims of all upon the fact of each man's
personal interest in his own happiness. Some more
recent sociological schools do indeed resume the appeal
to hedonism; but they do so—as we shall shortly note
—in connection with a doctrine of evolution which was
unknown to Comte, and which those who rely on it
regard as affording a new basis for morals, a new ram-
part against the assaults of a destructive individualism.
To unsophisticated phenomenalism, one fact is as good
as another; and there is no fact more pressing than the
claims of self. It may possibly be argued that the new
doctrines of evolution bridle the spirit of selfishness by
showing that each individual inherits a sort of com-
pendium of the moral experience of past ages. But, at
any rate, in the absence of evolutionary doctrine, Comte
had to qualify or corrupt his phenomenalism in the in-

terests of the public weal. It is not because experience
proves society to be the true source of individual happiness that Comte champions society, or that he sings the
praises of the social life. He ignores our specifically
human experience, and assimilates man's life, as far as
possible, to natural or animal existence. He will not
admit that reason has disintegrated the purely instinctive
co-operation of gregarious animals, so that it can never
be reconstituted. And he has no vision of a higher
fellowship, created only by the rational and moral nature
of man, or by that glorious Nature whose image is borne
by man alone, of all creatures upon earth. Comte has
his psychology of the rational nature,—of its characteristic selfishness and its no less characteristic unselfishness; but his doctrine, as we shall see, is profoundly
unsatisfactory, and his appeal to biology is a counsel
of despair. Instead of saying, "On to the fuller development of reason and goodness, for the cure of the
ills under which we groan," Comte says rather, "Back
to the life of sense, in which these ills had not yet
emerged." Comtism ignores the idiosyncrasy of man as
a rational being; hedonism at any rate recognises it in
however perverted a form. We must seek to attain
some worthier recognition of the great fact. Biology is
indeed a parable of the moral life, but still it is only
a parable. The resemblances are counterpoised by
immense differences. When these differences are
neglected an appeal to biology in the interest of morals
becomes a piece of mere improved assumption. And
Comte is more dependent on this appeal than he ever
clearly admits. He is more dependent on it than his
principles quite warrant. The only fashion in which
Comte is able to say "You ought" is in the formula,
"Society is an organism." Other sociologists have

other reasons for making the appeal to biology; it stands for this in Comte. And therefore this appeal in Comte is not a scientific statement of fact, but rather a rudimentary and defective form of the moral judgment,—valuable, no doubt, but valuable upon the principle which makes the one-eyed man king of the blind.

NOTE A. *On "Natural Law in the Spiritual World"*

[The appeal to biology has been traced in a different quarter, in the lamented Henry Drummond's first and brilliantly successful book, *Natural Law in the Spiritual World*. This book was criticised in a pamphlet under the name of *Biological Religion*, by one less widely known, but not less deeply loved in life or lamented in death, Dr. Finlayson.[1] Drummond of course appeals to the sharp modern doctrine of biogenesis, with its denial of all forms of spontaneous generation or xenogenesis; with its assertion of life from life, and like from like. It is certainly curious that an age which has taken stock so heavily in evolutionary speculations —and the very men of science who were pioneers in evolution and popularisers of its results—should also have reaffirmed, on the ground of fresh experiments, a view of life closely associated with creationist doctrines. Drummond, for one, appeals in his early work to biological science, because he is a theological creationist. His analogy is somewhat wire-drawn; his biology is of the simplest, rarely going beyond the single point named; when it does go further, as in discussing *Degeneration* or *Parasitism*, still extremely simple, and not very consistent with the foundation doctrine

---

[1] The late minister of Rusholme Congregational Church, Manchester.

of biogenesis. The book really offers us Neo-Calvinist religion, or even Neo-Gnostic, more truly than biological religion; but it shows the same contempt for metaphysics and the same blind confidence in empirical science which distinguish Comte and many lesser sceptics. Its religious teaching is often admirable, but the parable on which it is built misleads the author, because he supposes it to be more than a parable. Intellectually, the best feature in the book is the determination to trace continuity between different worlds of thought. This effort reappears in Drummond's later book, *The Ascent of Man*, of which we may have something to say hereafter. Otherwise, the later treatise is largely an inversion of the previous one. It obliterates the theological discontinuousness between the natural and the spiritual man, which had been so strangely supported by the assertion that the laws of physical nature must be viewed as continuous and operative in all regions of experience, even the most spiritual.]

CHAPTER IV

THE APPEAL TO HISTORY

In Dr. Hatch—Criticism—In Ritschl, how far Comtist—Other appeals; to *historic parallels*—Example from Comte—To the *whole tendency of history* —More usual in Comte; examples—Criticism—Mr. Mackenzie's criticism —Guidance to be gained from history is limited—Comte's varied and capricious appeals to it.

To appeal to history for guidance is a very natural resource on the part of those who distrust philosophy. It is found even among theologians who are interested, as Comte was not, in preserving belief in God. Probably the appeal was never made with more clearness or with more confidence than by Dr. Edwin Hatch in his St. Giles Lecture, "From Metaphysics to History."[1] Dr. Hatch can find no language in which to express his contempt for metaphysics, or his confidence in modern physical science. "We have passed into a new atmosphere. We have around us, not the glamour of a splendid mist, but the light of day." Science has "passed from metaphysics to fact, and" has "passed thereby from doubt to certainty." One province remains to be liberated—that of theology. Let us make a similar transition here, "from metaphysics to history"; then, even in theology, we shall find solid ground below our feet. The history which Dr. Hatch has in view is

[1] Published in the *Contemporary Review* for June 1889.

history of doctrine, the history of theological beliefs.
If we treat these in the light of the comparative method
they will no longer be meaningless, but meaningful; we
may even discover that "God is not only revealing
Himself to His creatures, but also realising Himself to
Himself" in history.

There is a great deal that is Comtist in this programme. To "abandon the search for essences and
look only to the operation of forces" is thoroughly
Comtist in spirit, though even "forces" is too metaphysical a term for Comte's taste; he would write
"sequences." The result contemplated, no doubt, is
anything but Comtist; but how immense the gulf between the method recommended and the results desired!
Either our Theistic beliefs are valid and defensible;
but, if so, there are other fields of knowledge besides
that cultivated by phenomenal science, and other
methods of study for metempirical and metaphysical
subjects. Or else Theism is merely a human delusion;
but, if so, historical science can do nothing to galvanise
it into fresh life. The sum of the longest series of
cyphers is still zero. In one thing Dr. Hatch is right.
Our age is pre-eminently an age of historical study.
Very likely our age does better work in dealing with
the history of beliefs, theological or other, than in
dealing directly with the problem of their justification.
Nay, our age may even make its best contributions to
metaphysics or theology at second hand in the regions
of history. But, if so, that is the weakness of our
age, not its strength. And, in any case, profitable
treatment of the history of such opinions implies a
belief that they deal with facts, not hallucinations.
Few of us, indeed, may be so metaphysical as Dr.
Hatch. A strange way surely of banishing metaphysics,

to propose construing God's realisation of Himself to Himself! The greatest idealists, with Hegel at their head, could not have improved on that programme.

Dr. Hatch appealed to the history of doctrine; it is in a different sense that the modern German theologians, to whom he stood nearest, make this appeal "from metaphysics to history." Ritschl and his school have mainly in view one race of mankind, and one epoch of time. They believe that, in the course of human history, truths have emerged and forces revealed themselves which satisfy human longings and lead human thought to its highest attainments. It is not merely history as a general survey of human development which they prize, but that history whose centre is Jesus Christ. Finding in history a revelation of Himself by God, they are able to honour history as the one true light of men. Otherwise unknown, God has here manifested Himself; otherwise unblessed, mankind here attains to happiness and salvation. Of course this sharply cut conception of revelation and its limits gives rise to very grave diffi- culties; but, amid all these, the appeal to history as urged by Ritschl has a seriousness and a significance which we cannot allow to Dr. Hatch's light-hearted paragraphs.

So far, there appears no kind of affinity between the Ritschl school and Comtism. Yet there are many symp- toms of relationship, and we find traces of them even in the matter now under discussion,—even in relation to the appeal to history. Much of the significance of Ritschl's appeal to history lies in the repudiation of the claim of physical science to rank as an authority in the spiritual life of man. Nature, according to Kaftan, is to be interpreted by history, not history by nature. As a progressive spiritual being, reaching his full stature under Christian influences, man claims that he shall not ultimately be made subject to the forces of blind and unprogressive nature ; he cries out for God to rescue the historical gains of human culture and human faith from the destructive forces of the natural world ; he finds God answering or anticipating his cry in Jesus Christ. There is nothing like this in Hatch. With him history scarcely differs from a new department of physical science. But we observe a manifest parallel between this Ritschlian position and Comte's subjective synthesis or subordination of the head to the heart. At the same time, there are immense differences. Justifiably or un-justifiably, the Ritschl school, amid all their scorn for dogmatic metaphysics, believe that they themselves, in their own way, have verified faith in God. They think that they have saved theology from the wreck of opinions, by stating it as a view of the contents of historical revelation, and as vouched for by its corre-spondence with man's nature and needs. In Comtism the subjective or affectional synthesis is admittedly a piece of human make-believe. Objectively corresponding to it, there is—nothing.

But how does Comtism itself, which has dismissed all interest in theology and all belief in God, make its own appeal to history for social guidance ? Or in what different ways may such an appeal be made, purely in the interests of society ?

The simplest view that can be taken is that which regards history as " philosophy teaching by examples." This view has been eagerly pressed upon our generation by one of its most brilliant teachers, Sir J. R. Seeley, though with a special reference to politics in the stricter sense, rather than to what we distinguish as social problems. Political history, according to Seeley, gives us the politics of the past, while present day politics are, to the statesman of wide views, history in the making. All manner of experiments in living, some of them successful and others unsuccessful, are recorded in the book of history. We moderns, with so immense a volume to study, ought to be safeguarded against many errors ; and we ought to find ourselves in possession of many pieces of practical wisdom, not as discoverers but as heirs.

Now Comte sometimes falls back upon the teaching of history in this simple and obvious sense. For example, he demands that the modern nation state should be broken up, under the positivist *régime* of the future, into fragments not much greater than the city states of antiquity. He allots to each a population of from one to three millions, the population of a great city, or of a canton or province of moderate dimensions. And he gives as his reason the teaching of experience, which is said to show that tyranny invariably sets in when larger aggregates are massed together in one political organisation. The assertion perhaps may startle us, but, true or false, it is an appeal to history, and an appeal to history in the obvious sense, in which history is regarded as a collection of examples or of experiments in living.

Oftener, however, Comte treats history in a different fashion. He would agree with J. S. Mill,[1] that, in contrast with the physical sciences, history discloses a law, not of repetition, but of continuous progressive development. Mill is careful to guard himself against making any assumption in this definition as to the moral value of one stage in history when compared with another. Progress in the moral sense he does not

[1] In his *Logic ;* and elsewhere.

affirm ; he affirms merely the technical law, that the
curve which describes the course of history never returns
upon itself. This belief is one of the characteristic
differences between the east and the west and between
antiquity and the modern world. The whole of
oriental mankind, with all its sages and all its faiths,
believes in the doctrine that history repeats itself.
It is part of the burden of the bitter book of Ecclesi-
astes in Old Testament Scripture ; after immense labour,
we find ourselves again exactly where we stood long
ago. Even in the West, the same doctrine was largely
held in classical times. Perhaps in the modern West—
in the Christian or semi-Christian West—we too easily
make the transition from asserting progress in the
intellectual sense, as a continuous evolution of change
from change, novelty from novelty, to asserting progress
in the moral sense, as continuous improvement.
Personally, no doubt, Mill himself believed in moral
progress as firmly as in continuous historical change.
And Comte believed both—the intellectual no less than
the moral ; " as if," he cries, " history ever repeated
itself." But, if history does not repeat itself, the past
cannot furnish examples to the present. If we are to
learn from the past it must be mainly in some other
way.

Shall we say then that we are to ascertain from
historical study which causes are gaining and which
declining ? And thereafter are we to shout with the
biggest crowd ? Is the teaching of history to be a
grandiose contribution to our study of the question
which way the cat jumps ? Comte's Law of the Three
Stages—an alleged continuous evolution in the history
of the past—may be so interpreted ; it may be taken as
a warning not to commit ourselves to modes of belief

which are plainly growing obsolete. And it may be
urged that, under due restrictions, there is high wisdom,
not ignoble policy, in bowing to the declared and
inevitable forces of history. Burke has given classical
utterance to this position in well-known words. "If a
great change is to be made in human affairs, the minds
of men will be fitted to it; the general opinions and
feelings will draw that way. Every fear, every hope
will forward it, and then they who persist in opposing
this mighty current in human affairs will appear rather
to resist the decrees of Providence itself than the mere
designs of men." It is this master current of tendency
which we are to think of as the *Zeitgeist*. The name is
not to be profaned, as one may say, by applying it to
every little ripple upon the surface of events. Mr.
Disraeli, presenting himself before the students of
Glasgow University as a wise and good old man, felt all
his wonted dramatic relish of the game of life in his
new part of Lord Rector, when he told his young hearers
that they must clearly understand the spirit of their
age ; perhaps they would feel themselves called to serve
it, perhaps to thwart it; but in any case it must be
understood. Such counsels assume that we mean by
the *Zeitgeist* paltry and sectional movements of mind.
But if we define the *Zeitgeist* in a limited and honorific
sense, resistance to the master principle of an age
comes perilously near to fighting against God.

In this sense some younger students of sociology
have deliberately suggested that one ought to learn from
history in what line things are moving, and then to help
the movement with all one's powers.

But here very grave difficulties suggest themselves.
If the unconscious reason of things knows in which
direction to move, presumably it also knows where to

stop, which is no less important. When the first railway tubular bridges were erected—the Britannia Bridge
over the Menai Straits, the Victoria Bridge at Montreal
—they were made much heavier than has been found
necessary in the light of fuller knowledge. What should
we say of the wiseacre who proposed to carry out the
principle of lightening railway bridges by constructing
them of lace or gossamer? In material affairs such
proposals are never made. One glance would show
their absurdity. But as mankind, especially in an age
of prevailing Agnosticism, stumble hither and thither
in search of social guidance, no absurdity is too crude to
find supporters; and many a tendency which was good
within limits is urged upon us without any limit
as the plain teaching of history. We have recently
emerged, or are emerging, from a period of emancipating
legislation, in which unwise or obsolete laws have been
abolished, and individual freedom has grown wider. The
tendency was doubtless good within limits; but does
this fact constitute any presumption whatever in favour
of the anarchist, revolutionary or philosophical, who bids
us entirely abolish organised government, and promises
in return a golden age of perfect happiness? The mere
fact that a policy was wise or was inevitable up till now
is no proof that it ought to be further persevered in.
The surgeon may have removed first a finger, then the
hand, then the forearm, as he found gangrene appearing
and reappearing; but that is no reason whatever for
operating at the shoulder if the upper arm is healthy.

Again, we may quote Mr. Mackenzie's statement of
the objections to the policy under discussion,—the policy
of pushing on along the lines where nature or history has
shown us the way. If we could be certain of distinguishing the master tendency of an age from the crowd of

rival tendencies in which it is all but lost, then history might be a sufficient guide. But too often, says Mr. Mackenzie, reflection becomes conscious of a social maxim only when the maxim is overripe, when it is ceasing to be healthy, or even to be completely alive. And so the conscientious student is apt to prolong the tendencies of the recent past rather than to detect the true needs of the present or the tendencies of the immediate future. He exhibits the weakness of the doctrinaire. The practical man, who is in touch with reality, though only half conscious of the principles and reasons *why* his policy is the right one, is more truly scientific than his pretentious critic in the arm-chair. When all men contribute to build a prophet's tomb, one may shrewdly conjecture that his message is no longer piercing and discomforting the conscience of the age. When impracticable politicians form a league for the defence, not of property only, but of liberty, one may fairly conclude that liberty is in no special danger, but that other interests of the commonwealth, not less vital to it, had best be looked to.

It would appear then that history cannot guide us very securely. It cannot guide us by quoting parallels from its repertory, for it is very hard to say what is a parallel ; and it cannot guide us by disclosing what is the master tendency of the present age, for such tendencies are seldom recognised in time. If history makes us wise, our wisdom arrives too generally after the event. Nevertheless, the study of history will be more and more imperative on all those who wish to counsel their fellows. It is mere waste of faculty to ignore the experience of the past, so far as that experience is available. Historical culture will give a man breadth of view. It will lead him to distrust sweeping generalisations and *a priori* formulas. It will teach him that every institution and method is relative to the social state of those by whom it is practised. But he who is to lead men strongly must draw wisdom from some other and higher source. History can give secondary elements of guidance; primary elements it cannot give. And there will always be the danger which that austerest of *libre penseurs* Mr. John Morley has emphasised, the danger that the historic method may justify anything in its own time, everything in its own place, and may relegate to limbo the distinction between right and wrong. Right and wrong—history illustrates that great polar contrast, but cannot fully teach it; yet after all is not that the beginning of wisdom? And is it not very nearly the end of wisdom too?

A last word must be added upon Comte's own use of the appeal to history, out of which so much of his sociological writing is composed. On the whole, he seems to owe a smaller definite debt to history than to biology. Sometimes he appeals to examples, as in the case quoted, when he refers tyranny to the undue size of the state. Sometimes he appeals to the past stream of tendency, as in his great generalisation of the three stages. Sometimes again he cuts right across the stream of manifest tendency; he surely does this in demanding that the large and organic modern state should be divided up into fragments; and in general no charge would seem to be more clearly made out than that Comte scarcely tries to show us his polity for the future growing out of the life of the past. Sometimes he appeals to a historical phenomenon, like the division of the spiritual and secular powers, which has struck his fancy. In such a case history is like a great magazine of wares, and Comte is like a purchaser strolling through

it, who puts down upon his list of household requirements—and Comte is catering for the household of
humanity—anything which pleases his own taste.   History is here the source of suggestions, and, as Comte has
much historical learning, he has a wealth of suggestions
at his command ; but history to him is certainly not a
ruler or a judge.   On the whole, Comte practises the
appeal to history with very little seriousness.   The
predominant partner in his lawgiving is the subjectivity
of Auguste Comte.

CHAPTER V

THE DOCTRINE OF ALTRUISM

A fragment of ethics—On a *psychological* basis—Opposes psychological hedonism —Healthily, but incompetently—Fitzjames Stephen's objection to it; we cannot alter nature's forces !—That is good determinism but bad morals— *Ethically*, is a new conception of virtue—Scientifically worthless [Mr. Baldwin]—"Balance" is preferred to altruism by Butler at times—By Spencer—Criticism.

A THIRD practical or moral authority is found by Comte in the doctrine of Altruism. *Vivre pour autrui* is to be our constant inspiration and our shining goal. This is really a fragment of that ethical portion of his system which Comte did not live to work out. The definition of Altruism is never formulated; it is never supported in argument; it is merely taken for granted. None the less it exerts an immense influence in Comte's own system, and has spread from it far and wide. Innumerable writers, Christian as well as non-Christian, have come to employ the term "Altruism" as a synonym for goodness. Such assumptions demand our scrutiny.

The doctrine has at least two aspects, a psychological and an ethical. Psychologically, it is assumed that human motives fall into two classes; one class terminating on the self, and seeking one's own private good; the second class terminating upon others, and seeking their good. It is further assumed that the division of motives into these two classes is exact and exhaustive.

The two classes in question nowhere overlap, and there is no third class of motives. Every action must be done with a view either to our own good or to the good of another, or some others, or all others. A further assumption is noteworthy, both psychologically and ethically. It is assumed that we are able, if we like, to encourage one class of motives and multiply the actions which proceed from it, to discourage the other class of motives, and to weed out or gradually exterminate the actions to which it gives rise. And, finally, there is the ethical assumption, that egoistic actions are bad *en masse*, and altruistic actions ethically good, so that plainly we ought to encourage altruism, and do our best to put down egoism.

Psychologically, this doctrine involves a notable break with the phenomenalist ethics of the past. Those systems had almost all been established upon psychological hedonism, on the assertion that man necessarily seeks his own pleasure, and cannot possibly, in any action, seek for any other end besides his own pleasure. Man, it was conceived, may be misinformed as to the best means of securing the given end, and therefore there is still room for ethical science as a body of prudential maxims ; it is still possible to say to man, hopelessly and incurably selfish as he is, " you ought " to do this or that ; although upon such a view " you ought " simply means, This will give you the greatest happiness in the long-run. Or hedonism might make room for ethics (of a sort) in a different fashion. The moral fellowship of human society might be regarded as a mutual insurance office, in which every one surrendered small fragments of present happiness in return for a guarantee against great contingent unhappiness in the future. Or by a sort of generous confusion the inference might be urged on men that, as each wants his own happiness, we must all labour for the happiness of all. But the psychological background of these various pieces of special pleading was the assertion that, first and last, each man seeks, and must seek, his own pleasure. The assertion can at times be made to appear almost self-evident, though a few minutes' handling by a skilled cross-examiner[1] will make it look very foolish indeed.

From that psychology to Comte's psychology, from old-fashioned phenomenalism to new-fashioned positivism, is a somewhat startling change. Shall we not welcome it as a change in the right direction? Certainly a less libellous account of human nature is given when we are told that it is composed of a group of selfish and a group of unselfish motives, than when the old view is reiterated, according to which human nature is root and branch, first and last, by eternal necessity, selfish and only selfish. But we must still inquire whether Comte's amended statement will pass muster scientifically, and, in the first place, psychologically. Now, Comte has no belief in a science of psychology. Psychology ought either to fall back upon physiology and phrenology, or to merge itself in sociology. Taken by itself, Comte regards it as a pseudo-science. But the neglected beauty has a capital opportunity for punishing the erring swain when Comte begins to talk psychology, for he talks nonsense. One may be confident of support from modern psychology in asserting that every action, however altruistic, is yet in some sense egoistic. It is *my* action. I should not have made the motive mine, it would not have moved me, unless I had found myself in its results. Mere altruism is mere irrelevance, the

---

[1] Cf. Prof. Sorley's *Ethics of Naturalism*, pp. 23, 24.

action of a lunatic, not of a sane man. Old-fashioned empiricism was right in looking for a personal motive in each action, though gravely in the wrong when it called that personal motive, uniformly and monotonously, by the name of pleasure. But again, with scarcely less confidence, one may assert that even the most egoistic actions are, in a sense, altruistic. Man is so radically social that his sins no less than his virtues are stamped with the signet of his nature. He sins socially. If he does not serve others he uses up others in his own service. Nay, even the cynic is only a social being in a pet. He retains the hope that some one is watching him. Diogenes, basking in his tub, has an exquisite pleasure in requesting the great Alexander to stand out of the light. Outwardly withdrawn from society, he is inwardly dependent on it; for admiration, or for criticism, but at any rate for notice. Of course, Comtists may rejoin that they mean to allow for all this. But does their formulation of the case satisfy the demands of science ? Surely Comte, of all men, will not maintain that scientific accuracy is superfluous, or that conduct can be safely guided in the light of slovenly and inaccurate thinking !

A second criticism is offered by Sir J. Fitzjames Stephen in *Liberty, Equality, Fraternity.*[1] Justice Stephen, like his brother Mr. Leslie Stephen, is a very severe critic of the weakness of Comte. He protests especially against a further assumption which we noticed in passing, the assumption that it is possible, by careful effort, to readjust the balance of egoism and altruism in human nature. According to Stephen, such a change lies as far beyond our power as a change in gravitation or magnetism, or any of the forces of nature. Sir Fitzjames

[1] p. 110.

Stephen does not (here at least) pin his faith to the old selfish psychology of hedonism. Allowing the assumption to pass, that there are a certain number of unselfish promptings in the nature of mankind, or of any given individual, he assumes that (like the elect under the scheme of Calvinism) they can neither be increased nor diminished in number. The criticism, advanced as it is by a determinist, is a very awkward criticism for his fellow-determinists to meet. Speaking as an impenitent freewiller, one admires this pretty quarrel between the forces of the enemy. Stephen appears to be the more logical or consistent determinist, while he is certainly the more impracticable and the more hopeless guide of human conduct. Put in so naked and outrageous a shape, determinism must repel all who love goodness better than they love paradox. Comte's determinism is disguised or kept in the background. He points out that human agency can do absolutely nothing to modify astronomical laws, but that, as we ascend the scale of the sciences, we see physical and chemical forces yielding more and more to human manipulation, until finally, arrived at sociology, we may well expect " the human providence " to prove itself nearly omnipotent. Stripped of its Comtist language, all this is true, but it is a truth incompatible with thoroughgoing phenomenalism. Just because man can modify nature, he can more profoundly modify himself. Just because he is not a passive stage, upon which the feelings fight out their battle and settle his destiny for him; just because " man is man, and master of his fate," he puts his mark upon the world in which he lives, and makes it *his* world.

We may now leave the psychological aspects of the doctrine of altruism, and consider its ethical aspects. It has been argued that the sharp contrast between

egoistic and altruistic actions or motives is vicious
psychology; and while we have agreed with Comte
against Stephen that the forces of human nature are
capable of being profoundly modified, we were sceptical
as to the possibility of harmonising this fact with the
principles of determinism. It remains to discuss the
ethical significance and trustworthiness of the altruistic
ideal.

Its significance in Comte's system is plain enough.
It furnishes him with a fresh definition of virtue, as the
appeal to biology furnished him with a fresh definition
of duty. Less authoritative than the doctrine of the
social organism, the doctrine of altruism appeals to
man's moral nature from a different side. To live
for self is αἰσχρόν; to live for others is καλὸν
κἀγαθόν. Thus there is a special appeal to *motive* in
this new definition. Perhaps, however, it is best
understood as a deliberate rejection of duty to God or
to any transcendent standard of worth. Virtue shall
be only barely mutual help between man and man.
Altruism accordingly is the religion of humanity itself,
considered as a law of conduct between individual and
individual. The state is not mentioned; society is not
formally invoked; but we are bidden live for others. It
is easy to see that this doctrine corresponds to a part, an
element, an aspect of human goodness. With Comte,
however, it stands for the whole.

The doctrine finds a response in human nature and
the human heart. For, whether recognised or ignored,
the moral nature of man is a constant factor in the
promulgation and the acceptance of ethical doctrines,
healthy or morbid. Conscience is always with us; it is
always more or less active, more or less influential; and
it sees something in "altruism." But, as a formal and

exhaustive definition of virtue, altruism claims to stand for everything.   And such a claim must be resolutely repelled.   If "altruism" were as clearly a psychological fact as it is (we believe) a psychological chimæra, yet, as a contribution to the science of ethics, it must fail.[1]

Badness is preferring myself to my neighbour; goodness is preferring my neighbour at the sacrifice of myself.   Yes, but what is that which it is morally good to bestow upon others?   Surely not the particular sensuous pleasure which I am forbidden to grasp hungrily on my own account?   If a man who drinks wine or beer in moderation gives up his own beer or wine that he may add it to the portion of his neighbour, and allow the latter to indulge a taste for drinking immoderately, that is highly altruistic behaviour, but it is not virtuous.   Indulgence may be as altruistic as any conduct whatever, yet indulgence is as vicious as any conduct whatever.

We need not wonder, therefore, if a further step is taken in criticism of such positions as Comte's.   Those who have discovered that we may sometimes do wrong in fostering the pleasure of others naturally go on to ask whether it may not be wrong to drop some of our own pleasures, or, at any rate, to drop some of our own rights?   Thus, in place of Comte's one-sided commendation of the service of others, we are asked to accept, as the true ethical ideal, a doctrine of balance between the claims of others and personal claims.   This conception— alternating, it is true, with other conceptions—is found as far back as Bishop Butler.   Butler has no very clear doctrine of the contents of the moral ideal.   That was

---

[1] Professor Baldwin (*Social and Ethical Interpretations in Mental Development*) seems to explode the contrast of egoism and altruism psychologically, and yet to take it for granted in ethics.

not the question which mainly interested him. When
he had said "Obey conscience," he thought he had
given the main instruction required of him as a moralist.
Still, the other question cannot be suppressed. Reasonable men must ask, "Granted that we are to obey
conscience, what is the general line of its commands?
What is the unifying principle of its various utterances? Surely it is incredible that such a principle
should be entirely lacking, and scarcely less so that
the principle of goodness should be inscrutable to a
reverent human inquiry!" Butler deals with this
further question, but he does so informally in a
series of not easily reconcilable *obiter dicta.*[1] Sometimes it seems as if benevolence were the master
principle of human conduct. In such passages Butler
takes his stand, where Comte afterwards rallied,
with the prophets of altruism. Sometimes, again,
Butler seems to speak as if conscience guided us just
where rational self-love would conduct us were it but
sufficiently far-seeing. In such words Butler conde-
scends to the cant, not of our century, but of his own,
though he does so with manifest uneasiness, and with
a bad grace. But, perhaps most frequently, he antici-
pates Herbert Spencer in pleading for a balance between
egoism and altruism. If we *must* define the principle
underlying good conduct, why, we find there are two
ultimate principles. At the back of our moral nature
there is, if not an irreducible multitude of special commands, yet an irreducible dualism—a pair of regnant
principles, and the line dividing them must be drawn by
a sort of practical tact. Theory is helpless to reach
past this "dual control."

[1] Cf. Dr. T. B. Kilpatrick's Introduction to Butler's *Three Sermons on
Human Nature.*

It is strange to find this doctrine of balance, this
glorifying of compromise, renewed by Herbert Spencer
—the second great name in the annals of sociology, the
inheritor of Comte's problems and Comte's vocabulary.
He also assumes the psychological legitimacy of the
contrast between " Egoism " and " Altruism "; but
altruism does not rank with him as a compend of all the
virtues.   It is only one half of virtue, though possibly,
in the language of children, " the biggest half."

Here again, as formerly, we have to ask, Which is
the juster development of the view in question ?   If we
accept altruism as a conception which is psychologically
valid and ethically important, ought we, like Comte, to
press it as hard as we can, or rather, like Spencer, to
urge that altruism is good only when balanced by a
judicious regard to our own egoistic rights ?   Perhaps
the latter view has more of the remnants of wisdom in
it.   But the truth is, both views are impracticable ;
Spencer's no less than Comte's ; a doctrine of balance
no less than a doctrine which ignores self.   The double-
minded man is, and remains, unstable.   It is impossible
to serve two masters.   A true moral analysis must
recognise something higher in the lowliest duty, and in
the commonest act of kindness, than private convenience,
whether that of *ego* or *alter*.   " One person I have to
make good—myself.   My duty to my neighbour is
much more nearly expressed by saying that I have to
make him happy—if I may."[1]   Yes indeed ; but, in
making my neighbour happy, I make myself good ; or,
if I fail to make myself good, I shall not long make my
neighbour happy.   Both are duties ; or rather both are
aspects of the good life, in whose unity they are merged.
And in both alike there is a reference to something

[1] R. L. Stevenson, *A Christmas Sermon.*

higher,—call it duty ; call it God's will.   In faithfulness
to one's own moral vocation, social and spiritual—in
faithfulness to "my station and its duties," primarily
and literally in the kingdom of Great Britain, but, by
ultimate analysis, in that better kingdom which cannot
be moved,—one is delivered from the extravagances of
altruism, and from the imbecilities of compromise, into
the very peace of God.

Seeing that men are quite sufficiently selfish, Comte's
rhetoric in praise of altruism has probably done little
harm.   As rhetoric, it is passable ; as a rough piece of
popular pleading, it will serve.   But it is wholly lacking
in the scientific quality which we were promised.   In
other words, it is destitute of exactness, or, one might
even say, of truth.

CHAPTER VI

COMTE'S LAWGIVING

Its principles—The separation of the temporal and spiritual powers—*Political character* of Comte's *sociology*—Details—Summary.

IT is not possible for us to give a detailed sketch of the *Positive Polity*. One can only notice in the briefest fashion how the superstructure answers to the foundation laid, or how the threads that have caught our attention are intertwined in the pattern of the finished fabric.

We have noted already the following points : the law of the three stages, or the alleged movement from superstition to science ; the movement from militarism to industrialism ; the separation of the spiritual and the temporal powers ; and the restriction placed on the size of states.

The third of these may need a word or two of explanation or comment. Under Positivism the separation of the spiritual and temporal powers is very much a separation between men of theory and men of action. By means of such a separation each class is to develop its own especial excellences, and the theories will be disinterested, while the practice will be—what? it is hard to say, perhaps more perfectly expert. Surely if any proposal deserves Comte's favourite reproach of "pedantry" this proposal deserves to be so stigmatised. It is a singular example of his fondness for "Catholicism *minus* Christianity." The director of conscience is to be made supreme in the whole life of Positivism.

The only general observation that need be added is upon the name Polity. Yes, it is indeed a scheme of *politics* that Comte has given us. There is no contrast left between the organised life of society and its more strictly sociological aspects as a natural growth. May we not say with Mill that the natural tendency of things is simply set aside? That no serious effort is made to show that the predicted future has its roots in the past? May we not repeat our previous statement that the predominant partner in the *Polity* is neither scientific biology nor scientific history, but the wilful will of Auguste Comte?

Some of the details of Comte's scheme may now be run over.

The business of government is to be assigned to a triumvirate of bankers, who are to act as dictators, after consultation with the "supreme pontiff," or head of the spiritual power for all mankind. No more is to be heard of popular rights, they are a metaphysical figment. Henceforth men are to speak only of duties. The dictators will accordingly name their own successors. Limitations on the powers of the dictators will nevertheless exist. *First*, there will be absolutely free criticism, at the risk of religious excommunication, or boycott, at the bidding of the priests. *Secondly*, the priests will act as a counterpoise; or rather the spiritual power will do so, composed of the priests *plus* the women *plus* the "proletariate."

The clergy are to be intellectual and moral experts,

living on salaries—small salaries. As we know already, they are to be debarred from political power and from business activity. The intellectual training of youth is to be entrusted to them, and also medical practice. They are, above all things, to beware of specialism. It has been remarked that Comte is almost as much opposed to specialist "pedantry" as to metaphysics. *The great champion of scientific certainty is becoming more and more jealous of mere knowledge.* Utility is to be everywhere kept in view. Priests are to "direct" consciences by counsel, not by force. It will be remembered that they may sometimes advise the dictators, and that, where necessary, they are to oppose them. The tremendous weapon of excommunication is in the priests' hands.

Business is to be carried on by captains of industry, directing *proletaires*. But capitalists who have had the benefit of positivist training in youth, and who walk all their days in the fear and love of the " spiritual power," are sure to regard their position mainly as a social function, and to seek for no profits beyond a reasonable salary or " living wage." If necessary, strikes and locks-out may still be resorted to ; but such an emergency can seldom or never arise, under the fostering care of a wise priesthood. Every man is to be regarded as doing social service by his work. No mere " cash nexus " is to hold society together. As with one's professional attendant, so with the tradesman or artizan whom one employs, one is to feel that he has earned a debt of friendship. On this point Comte's teaching is surely large-hearted and nobly wise.

Positivist education, especially as carried on by mothers, will be moral even more than intellectual. And afterwards, the influence of the priesthood, of public opinion, of the boycott, and of some other institutions of positivist religion, will help altruism to gain the mastery.

Religion consists chiefly in prayer, offered morning, noon, and night, and addressed to humanity, especially as represented by one's female relatives—mother, wife, and daughter. If any one is lacking in the second or third of these, or if any one's wife or daughter is inadequate to the *rôle* of representing humanity, one may substitute other ladies in one's mind. Humanity consists of the *good* alone—the good of the past, the present, and the future—along with those races of the lower animals which, being specially serviceable to mankind, are " incorporated in humanity." A calendar of saints' days helps to keep the great names of the past in remembrance. For one's own part, one may look forward to something of a similar " subjective " immortality. Along with humanity, the " great being," the earth may be worshipped as the " great fetish," and space as the " great medium "—together constituting a Positivist Trinity. Paris will be the spiritual capital of humanity. Auguste Comte is the first pontiff of the new and definitive form of religion,—a distinction which is no more than fitting in the case of one who combined in his own person the merits of " Aristotle and St. Paul." —Comte admired Aristotle as heartily as he disliked Plato, and he went far beyond Tübingen itself in styling St. Paul "the real founder of Catholicism," *i.e.* of Christianity.

He not only fixed all these matters, he fixed innumerable others. Every man of business was to retire at the age of sixty-three, spending the remainder of his days in advising his son how to carry on the business. Every labourer was to own the house he lived in. Every house was to contain seven rooms—no more, no less. The labourer's salary was to be 100 francs for a month of twenty-eight days, or an equivalent calculated in piece wages. Every treatise was to contain seven chapters, each divided into three parts, each part sub-divided into seven sections. Every poem was to contain thirteen cantos, thirteen being another of Comte's sacred numbers.

But, *Quousque tandem?* Have we not had enough of *this* version of scientific sociology ? In point of fact, we find ourselves, under Comte's guidance, in a world of caprice. Biology gives him a parable of moral truth, not a law ; history offers suggestions to the philosopher, but does not control his judgment ; the ideal of altruism, of which he is the prophet, is an unproved and unsafe assumption. A brilliant and erratic man, he rode his hobbies hard, and threw the reins upon the neck of his fancy as he approached the details of conduct. If science is definite, measured, certain in its utterances, then Comte, in spite of his aspirations, is no true scientific leader for the human race.

PART II

SIMPLE EVOLUTIONISM—SPENCER, STEPHEN

CHAPTER VII

DARWINIAN AND SPENCERIAN CONCEPTIONS OF EVOLUTION—DARWIN

Evolution came as a surprise—Darwin deals with biology—With species only—Taking "Struggle" from Malthus, he perceives in it (Natural) "*Selection*" —A true cause, but minute; an immensely slow process—Compare the replies to Malthus—*Sexual Selection* accelerating—Or *Use-Inheritance*—But too much Lamarck, making variation not "casual," but purposeful, would render unnecessary the "Selective" action of "Nature"—Recent doubts as to use-inheritance.

THE appeal to biology, so far as it was formulated by Comte in the interests of social science, did not seem to possess any great significance. The immense rise in importance that was to accrue to biology from the evolutionary theories of this age was hidden even from the best minds of the preceding age. Even Hegel speaks scornfully of the foolishness of trying to read the purely ideal evolution, described in his system, as a process in time; but those who feel his influence most strongly to-day have generally accepted the identification. Comte goes further still. He expressly names hypotheses regarding the origin of species among the wasteful and unprofitable inquiries which the human providence will discourage and put down. So unfit are even the learned to play the part of providence. So liable are they to misjudge doctrines which, even if destined at last to be regarded as one-sided and more or less fallacious, have yet shown themselves immensely fruitful in suggestions bearing upon every branch of human knowledge. It is now admitted by able adherents of Comte's system [1] that the doctrine of evolution supplies a background or basis for Comte's unification of knowledge. In such a statement Spencer's form of evolutionary doctrine seems to be most directly contemplated, and Spencer is perhaps the least thoroughly biological of all the evolutionary thinkers, whether moralists or sociologists, whom we shall have to pass in review. Yet the great movement of our day was in connection with a biological doctrine which Spencer will certainly not repudiate. And it falls to us rather to argue for a difference than for a kinship between Spencer and Darwin. The kinship is claimed, asserted, conceded.[2] We do not deny it; but we believe that the differences reach deep down. Before we go further we must take a hurried view of evolution as conceived by both these influential writers —and first, as conceived by Darwin.

Darwin's problem, vast as it was, and bold as was the effort required to deal with it, was strictly limited. It lay within the world of organic life. It sought to account for the origin of distinct species among plants and animals. Organic evolution, as taught by Darwin,

[1] *e.g.* Mr. J. C. Oliphant in *Chambers's Encyclopædia*, 9th edition.

[2] Mr. C. W. Williams, of whom Mr. Spencer complains, certainly seems to underrate Spencer's originality (in comparison with Darwin) upon p. 2 of his *Evolutional Ethics;* but he makes concessions on the other side upon p. 28. Our desire is to show that the two great men moved on different lines.

means, one takes it, the evolution of organisms, a doctrine
of evolution *versus* (special) creation as accounting for
species, though the phrase organic evolution is sometimes perhaps used by other writers[1] in a wider, or
vaguer, or deeper significance. Darwin himself, as a
specialist, had nothing to say to us on the origin of life,
nothing, assuredly, on the origin of the universe. At
one point, indeed, he unavoidably opened up very deep
problems. For among the species with which he dealt
was the human race ; and a discussion of the origin of
mind involves a reference to the beginnings and ends of
all things ; it forces us back to first principles and
drives us on to the final problems. But of this, perhaps,
Darwin was never adequately aware. Every one who
has studied philosophy sees it, but Darwin, though a
specialist of genius, and a specialist on a great scale, was
still, after all, a specialist. And he never claimed to
bring the world a new cosmical philosophy ; it was
enough for him to introduce one new hypothesis, linking
together all forms of life, and to see this hypothesis
conquering mind after mind, until the whole civilised
world seemed to bow to its discoverer. Darwin dealt
with the evolution of species, Spencer has dealt with
the evolution of the universe.

What, again, was the special contribution made by
Darwin to his problem—so old a problem, with which
so many minds had grappled, and, on the whole, so very
unsuccessfully ? Primarily of course it was the doctrine
of natural selection through the struggle for existence.
As students of social philosophy, we are specially interested to recall that Malthus's doctrine of population
directed Darwin's attention to the aspect of struggle in

[1] *e.g.* Dr. E. Caird. In a *deeper* significance, perhaps, as implying
necessary or organic relation between the organism and its environment.

nature, a fact or aspect of things which he speedily traced throughout all living nature, vegetable or animal. But the doctrine of natural selection—of survival of the fittest [1]—of improvement of species through the struggle, and gradual development of new species—that was Darwin's own brilliant corollary. He perceived that selection was sure to accompany struggle, if at least there were any differences or variations separating competitors from each other. The best man, or brute, or plant must win, upon the average, and in the long-run, if only there were better and worse, better and best, blended in the competition. Otherwise struggle might mean deadlock and mutual exhaustion, as of two equally matched armies after a long campaign, and general doom to extinction, as of the survivors from a wreck when food runs short. But variations do notoriously exist. Nature, which, "red in tooth and claw," unmistakably asserts the fact of struggle, not less clearly reveals the fact of selection with its two sides of defeat and victory, and with its basis in a tendency to vary. This variation is mainly conceived as congenital. Some are born better, some worse. Not only are the offspring of better parents better equipped; within the same family, as experience shows, some are better equipped than the rest, some sink below the average. How far this tendency to vary went, Darwin never dogmatically affirmed. It was enough for him usually to treat it as casual and therefore as undefined. The great concern of nature, the arch examiner, was not to secure good candidates, but to secure a plentiful flow. If there were but enough, some good specimens would assuredly be found. So said, so done; *teeming* nature, as we call it, brought forth all things abundantly, ay, and superabundantly; not

[1] Spencer's phrase, however.

monotonously, in mechanical batches, but with minute
yet important differences ; the result was continuous
adjustment, and adaptation, and evolution, and improvement, at the cost of a heavy and remorseless "pluck,"
year after year, age after age.   Finally, what variation,
and struggle, and selection have beaten out, heredity
preserves.   Within the limits of variation heredity
perpetuates, in the offspring, the good and victorious
qualities of the parents.

This, in very rough and brief outline, is the central
portion of Darwin's hypothesis,—the doctrine of natural
selection through struggle.   When this doctrine is
applied to morals or politics, we have Darwinism in
morals or politics.   Where this doctrine is absent or
subordinate, we may have evolutionism in morals or
politics ; Darwinism we have not.   In this lay Darwin's
superiority over many evolutionist predecessors, he had
laid his finger upon a *vera causa*, an undeniable fact in
nature,—the abundance of offspring, or — otherwise
roughly stated—the scantiness of food ; upon an undeniable tendency in nature ; a tendency to improve and
modify all living forms,—improving them, *i.e.*, so far as
to make them fitter to survive in their given environment.   Theories like Lamarck's of the direct action of
environment might be plausible, but they seemed to
lack verification.   Darwin's theory sprang into a different
position because it appealed undeniably to real facts ;
although it gave them a very startling extension in the
range of their operation.   Certainly the plain man would
have said that the tendency, though real, was too
infinitesimal for its work.   One would have said that
natural selection was as utterly unable to explain
variety of species, as Sadler's doctrine, or Herbert
Spencer's hope, to meet the difficulties alleged by

Malthusianism regarding the human race. No doubt, human reproduction becomes less rapid as population thickens. The alleged self-correcting tendency of the growth of population is a true cause, so far as it goes; or rather it is a group of causes, urgently requiring to be disentangled, to be studied, named, estimated one by one; but, in their whole result, they are altogether insufficient to check over-population. And in like manner Spencer's cause is a true cause. It is undoubtedly true that there is a general correlation of fecundity with a low position on the evolutionary scale; it is true that, as mental and æsthetic interests count for more, the physical tendencies of sex will count for less in the human race; yet, as far ahead as we can trace, there will still be problems of population. So one would have said of natural selection too: It is a true cause, but cannot possibly do the work asked of it. Its effects are minute; being minute, they will be immensely slow in achieving anything. A blind and indirect method of selection, by striking out all the unfit—by trial and error—is the most tedious method possible. If at every cross-roads I have to follow each track in turn, taking them as they come, going on in each case to the next town before I can learn whether I am on the right road,—if I am wrong, coming back from the town to my cross-roads and trying the next track till I find a town upon it, and so forth and so forth — plainly, it may take me all my days to work my way to my chosen destination.

Darwin's theory, however, includes other elements besides natural selection; and these, if reliable, seem to point to agencies which would accelerate the process of evolution. One addition which Darwin proposed to his doctrine was sexual selection. "None but the brave

deserves the fair "—that is half the new doctrine. For
sexual selection is believed to exist in two forms ; first,
when the males fight with each other for the privilege of
access to the females, as in the case of lions or stags :
secondly, when the males vie with each other in æsthetic
attractiveness, as Darwin supposed to be the case with
birds, and as a larger number of observers believe to be
demonstrated in the case of certain insects. The assumption appears to be that the unsuccessful males remain
almost or altogether sterile by force of circumstances ;
accordingly, a criticism passed by Wallace upon Darwin's
theory of a sexual selection *in the case of birds* is to the
effect that, apparently, even the least beautiful of male
birds finds a mate sooner or later during the pairing
season ; that the inferior forms leave offspring as well as
the superior forms ; that accordingly no selection between
different forms is due to the imperfect rivalries of court-
ship. It might be possible, surely, to meet even this
difficulty. Presumably, the successful males, whether
fighters or beauties, will pair off with the most desirable
females ; there will be an intensified divergence of
offspring in the next generation, with consequent
emphasis upon variation, and hastening of the final
victory of the strong over the weak. On the other
hand it may be held that sexual selection—in this sense
—is only a remedy for an obvious weakness in the
process of natural selection,—the danger that advantages
will be lost by crossing. But if, as is usually thought,
sexual competition implies the celibacy or nearly so
of the unsuccessful candidates ; then we have before
us a direct and psychical process of selection, not an
indirect and natural process ; a short and straight
process therefore, not a long and circuitous one. Of
course, one is not guilty of the absurdity of saying,

that the females are conscious of a preference for the best male specimens *qua* best, or are urged by an enthusiasm for the ideal! We only affirm that, in virtue of their animal minds, they yield themselves to the stronger or to the fairer. Yet again a question may be raised, whether the evolution of beauty, supposed to enter into the second form of sexual selection, is necessarily the same thing as an evolution in strength and efficiency. It may well be so. Beauty may well be correlated to those qualities of health and vigour which make a type intrinsically fitted to survive. As Mr. Grant Allen once remarked in a rare moment of inspiration or common sense, the saying that beauty is only skin-deep is itself but a piece of skin-deep and superficial wisdom. Yet, even if beauty does not imply superior health and vigour, so long as beauty is not developed at the sacrifice of useful qualities, sexual selection will hasten the evolutionary process along lines on which it has already begun to move — along the line of beauty, if not incontestably along the line of strength or aggregate fitness.

Another supplement to Darwin's central doctrine is what may conveniently be termed use-inheritance. This played a great part in the evolutionary theories of Lamarck, along with a still more questionable doctrine, that of direct adjustment of the organism to its environment. As the comic song puts it, the giraffe got a long neck by stretching to reach the upper branches. That is scarcely Darwinism; it is much nearer Lamarckism. The Darwinian giraffe happened to be born with a longer neck than the remainder of his family, and consequently outlived them all in a time of scarcity, and was the only giraffe who transmitted his qualities to offspring.

If the giraffe stretched its muscles and its vertebræ to their utmost, and begat a son whose neck, unstretched, was as tall as the parent's in his habitual tiptoe attitude, that would be use-inheritance—one-half of Lamarck's doctrine, and an accredited though a subordinate portion of Darwin's. If, however, the hungry giraffe organised in itself by some means or other an extra joint, or an extra set of muscles, or, as would probably be necessary, both, that would be a grotesque illustration of the second half of Lamarck's theory,—of direct action by environment in the way of modifying an organism ; a grotesque illustration of a sufficiently grotesque belief. At times, it is said, Darwin writes as if he were willing to admit this, viz. as a source of variations. But he has never formulated a theory of the cause of variations. He is content, as we observed, to treat them as casual. That, however, cannot mean that they are uncaused, or that the uniformity of nature breaks down as we approach microscopic cell processes. Perhaps at the utmost we can justify the phrase by taking it to mean that congenital variations from the parental qualities are neither on the average advantageous to the species, which might be repudiated as a somewhat strong teleological doctrine, nor yet disadvantageous to the species, a view which would imply a sort of dysteleology, as if we lived in the devil's world, and evolution had to go on with a dead heave in spite of the recalcitrance of nature. Chance or accident in common language means "not purposed," and it may perhaps be fair to call variations "casual," if they stand on the average neutral to the purpose or end of the species, viz. to survive and propagate itself. Still the epithet used without analysis is rather slovenly, and any thinking which is fairly summarised by the use of that epithet must be regarded as rather slovenly too.

Or, if we hesitate to say this of Darwin, we may at least affirm that he left much ground for subsequent investigation. He concerned himself but little with the laws determining variation. There *were* variations; there *were* candidates of varying degrees of merit. Get me candidates, he said in effect; I will give you an examiner who, however tedious in method, is in the long-run unerringly wise. Nature will select, come the variations how they may. At times, as we have said, Darwin seems willing to accept Lamarck's cruder and less verified doctrine, of a direct self-adjustment of the organism to the environment as a source of variation. Plainly, however, if this does occur, then, so far as it occurs, it supersedes natural selection. The supplement to the theory will displace the theory itself. Those called in to give help as allies will remain as absolute sovereigns. There is no need of indirect methods for compassing a teleological result, if such a result may come about directly through the living powers of the organism. We shall do well then to neglect this admission by Darwin in favour of extreme Lamarckism, particularly as it seems to be a mere *obiter dictum.*[1]

Even use-inheritance, however, will avail to shorten the process of natural selection. The offspring will start at the point which the parents had reached when it was conceived, not at the point where the parents themselves started, nor yet at that point *plus* a certain

---

[1] Darwin's clearest references to the causes of variation are probably found in his *Variation of Plants and Animals under Domestication.* The theme is therefore a restricted one, and it must be added that the language employed is less clear than would be wished. The following references may be consulted : vol. ii. pp. 290, 305, 311, 552. It should be added that to a certain extent *any* reliance on Lamarckian factors, even for " use-inheritance," tends to throw the tedious process of natural selection into the background.

amount of casual variation. On the other hand, we shall have to notice later on that this accelerating process of use-inheritance is much less confidently believed in to-day than in the hour of Darwin's absolute supremacy.

CHAPTER VIII

DARWINIAN AND SPENCERIAN CONCEPTIONS OF
EVOLUTION—SPENCER

A cosmic philosophy—Resting on correlation of forces—And on hypothesis of organic evolution — Emphasising natural (physical, material) law — Darwinism as a cosmic philosophy? Alexander—Cf. Lotze—Cf. Fiske— Spencer values true use-inheritance as accounting for *a priori* knowledge —But natural selection is *not* the source of his *laissez faire* doctrine ; he looks forward to a future "balance"—His relation to embryology—*Evolution* means growing complexity—In terms of matter—Two other phases— *Dissolution* as death—As catastrophe—*Equilibrium* is theoretical and prophetic—Spencer's sequence of the three phases—Criticisms : on the assumed *beginning* of the process—On its *isolation*—On *equilibrium*, as involving a different point of view—Reason is more than a new phase of complexity—The whole process breaks up into a series of separate evolutions in complexity.

MR. H. SPENCER's problem is wider than Darwin's, extending, as it does, to the whole of the phenomenal or "knowable" universe. The impulse to it came from two scientific theories of the age. The first was Grove's proof of the correlation of the physical forces, clenched by Joule's determination of the mechanical equivalent for heat. As a result of this, the inorganic world seemed to gather itself together in one, and to manifest its unity as it had never done before. Phenomenally, the differences remained ; heat was heat, light was light, electricity was electricity ; but it was now proved that some were mutually convertible, and it was henceforth probable that all were so ; it was known that some were modes of motion, and it came to be believed with increasing definiteness that all the others were equally modes

of motion.   In the invisible world of molecular change
it was assumed that these diverse branches combined in
one common trunk.   The second discovery was Darwin's
account of the origin of species.   Before this theory was
broached Spencer was already on the track of his own
thoughts.   If it helped him it did so rather by confirm-
ing his original bias than by making him a convert to
the special peculiarities of Darwinism.   In its simplest
shape Spencerian evolution is an assertion of the all-
sufficiency of natural law, a denial of intervention from
outside at any stage in the process by which the universe
has become what it is.   Moreover, natural law means
here strictly physical law ; everything is to be explained
in terms of " matter and motion."   This denial of all
miracle, and of everything analogous to miracle, gives
evolution its charm in the eyes of a fighting evolutionist
like Mr. Edward Clodd.   On Spencer's premises " there
is nowhere else " outside the process whence interference
might come.   Mr. Spencer is confident that he can
account for the beginning of the whole process.   The
inorganic world has been unified by one discovery, the
organic by another.   True, the transition from one to
the other had not yet been cleared up in terms of natural
law ; nor has that been done, one may add, until this
day ; but by an act of scientific faith Spencer affirms
that the last remaining gap must also be filled up, and
natural law remain as the power from which all things
have proceeded—master of the whole situation.

When we ask whether there is any close connection
between Spencer's philosophy and the doctrine of
struggle for existence, we feel at once that Darwinism
is almost impossible as a cosmic philosophy.   Professor
Alexander seems, indeed, to contemplate giving a
position of universal importance to the Darwinian

doctrine when he writes as follows : " The application of evolution to morals may mean only the employment of biological ideas ; or it may mean that morals must be treated as one part of a comprehensive view of the universe, in which a steady development may be observed from the lowest to the highest phenomena, *and a development, it may be added, which follows the law of the survival of the fittest.*"[1] The use of biological ideas we have seen in Comte, though doubtless only in one of many possible applications. We shall not find much more in Mr. Leslie Stephen's *Ethics*, though he has of course, in the background, a belief in evolution on the grand scale, as a cosmic philosophy. Spencer works out such a philosophy, and we see in it a considerable amount of pressure directed upon ethics from other parts of the fabric of knowledge. But in Spencer there is no attempt to take the law of the survival of the fittest out of its biological limits, and to give it a cosmic significance. So far as he traces an influence from one cosmic system upon another which has advanced any distance along the evolutionary path, he regards such influence as purely mischievous. It makes for dissolution, but not for evolution. Perhaps even Mr. Alexander did not seriously mean to include the physical " universe " in his Darwinian scheme. Competing organisms we know ; are competing universes anything better than a delirious dream ? Organisms die out, not because they are too ill-balanced for the tasks of life, but because they are, on the whole, in their own environment, inferior to other organisms, and therefore succumb in the competition. We must go back to very early " pioneers of evolution "—to Democritus or Empedocles—if we are to find survival of the fittest

---

[1] *Moral Order and Progress*, p. 14.

seriously applied to the cosmic process.   Yet its logical possibility is pressed upon us by so distinguished a man of science, philosopher, and theist as Hermann Lotze. " With reference to the past, we are at liberty to assume that at first an innumerable multitude of inharmonious forms, intrinsically hostile to any end, actually emerged from the reciprocal impact of blind elements; that these forms, however, were not able to maintain themselves in the course of nature, as against the contrary assaults from without; that on the contrary only those few held out which had chanced to be the more fortunate; that then these fortunate ones exerted more and more a determining influence upon the rest; and that thus gradually it has come to pass that nature runs its course, not indeed in complete and perfect conformity to an end, but after all to such an extent that there still remain but few disturbances or interferences by which the development and perpetuation of the structures that are conformable to an end is endangered.   In this way, therefore, it would not be unthinkable that an original chaos gradually shaped itself into a nature that is arranged in conformity to ends." [1]   Moreover, the postulate underlying such a view—in Lotze's opinion, of course, a mere logical abstract possibility; in no wise a fact—is given on the previous page : " If we take for granted that an indefinite multitude of different elements act upon one another entirely in accordance with mechanical laws, and that they were aboriginally in reciprocal motions, which were not regulated by any design."   This postulate, named by Lotze only that he may presently dismiss it as metaphysically untenable,[2] is

---

[1] *Outlines of Philosophy of Religion*, tr. p. 20.

[2] The many elements reducing themselves to elements in one great system ; the separate processes to one many-sided evolution.

identical, not perhaps with Spencer's, but certainly with his disciple Fiske's, " the mere coexistence of innumerable discrete bodies in the universe, exerting attractive and repulsive forces upon each other." [1] Spencer, perhaps characteristically, prefers to give us vague glimpses of a " homogeneous " though highly " unstable " continuum in space, finite in its dimensions, as the origin of all change. We conclude, therefore ; a cosmic philosophy might perhaps be grounded on a more than Darwinian apotheosis of competition. But no modern has tried to work out such a scheme—unless Lotze in one of his paradoxical moods as the candid friend of theism. Fiske might have been tempted in that direction, but was not. Spencer did not even cast one glance towards it.

Only one part of Darwin's theories is specially important to Spencer—the Lamarckian doctrine of use-inheritance. That is the basis of Spencer's reconciliation of Intuitionalism with Empiricism. We modern men possess intuitive knowledge—partly of mathematical, partly of moral truth—simply because our ancestors have had a wide range of experience of mathematical and moral facts, and have been able to impart their principles to us in the shape of innate tendencies to believe—tendencies which forestall experience and anticipate its results ; generally with accuracy. Thus Spencer has an answer for many difficulties. What gives conscience its awful authority over the human spirit ? What makes right and wrong so different, psychologically, from a calculation of consequences ? Why, the experience of law-abiding and dutiful generations, whose blood flows in your veins. Again one asks ; what is the hold that the public weal

[1] *Cosmic Philosophy*, ii. p. 867.

has upon me, a separate individual, with my own desires, ay, and my own rights? But his reply is
ready. The tribal or national conscience is within
you; it is a part of you from your birth; sinning
against it you sin against what is best in yourself.
Morally, however, Spencer gives this no great range,
and his colleague or disciple, Mr. Leslie Stephen,
writes a treatise on ethics without once mention-
ing it. Spencer is little inclined to admit true moral
axioms; he is resolved to keep the door open for
a phenomenalist doctrine of "causal connexions" in
conduct, if not exactly for hedonistic sophistications.
It is elsewhere that he has frankly confessed the existence of axioms, mathematical or "transcendental." He
has got his explanation of these, if he is allowed the
appeal to use-inheritance; but if not! Spencer is
fighting for his hearth and home, and for all that he
counts most sacred, when he girds himself to refute
Weismannism off the face of the earth. Apart from
use-inheritance, indeed, one does not see how the
evolution of mind is ever to be made decently intelligible, unless because "intelligence" was in the
beginning a "casual variation" of small amount—and
the stupider specimens died out, etc., etc.! That
explanation will never fail those whom it can satisfy.

Except on this point of use-inheritance, Spencer is
hardly to be regarded as Darwinian in his thinking.
Natural selection has hardly influenced his statement. I
do not mean that he refuses help from the doctrine, when
he finds help offered incidentally, in the biological or historical region. He is too good a tactician to do that.
But Professor D. G. Ritchie seems quite unwarranted in
explaining Spencer's *laissez faire* individualism by his
bigoted attachment to the doctrine of natural selection

by struggle. Far from that; Spencer's golden age of individualism lies in the future, in a period of equilibrium; but if struggle is all-important, such a period can never arise. Over against Darwin's conception of many organisms competing with each other, Spencer sets up a picture of one great peaceful process. Mr. Leslie Stephen tells us we ought perhaps to regard humanity as a single organism; Spencer seems almost to regard the whole of the universe as one great organic growth. Embryology shows him the simple almost homogeneous cell differentiating itself and growing complex; it is the same process Spencer traces in the universe, though he states it in terms barely of "matter and motion." [1]

What then is evolution, that key to the whole knowable universe, as stated in Spencer's own system? What are its great laws, or what are the properties manifested by "matter and motion" as the subjects of evolutionary change?

There is one word which may state sufficiently for our purposes what is meant in Spencer by evolution—the word complexity. Evolution means growing complexity; more complex is more evolved. Whatever technicalities are unfolded in the successive definitions given in the course of the volume upon *First Principles*, they do not carry us beyond this contrast of the simple and the complex. They are drawn up "in terms of matter and motion," which means that the details of the definitions apply to inorganic matter or to the physical basis of life, but cease to bear any meaning in psychology and sociology, in what Mr. Spencer calls "superorganic" evolution. It may plausibly be held

[1] Spencer has admitted his indebtedness to von Baer the embryologist for the idea to which he has given so wide an extension.

that, as knowledge advances, thought grows continually more complex, though it may be questioned with something more than plausibility whether it is possible in ultimate analysis to resolve the complex of consciousness into isolated presentations—even if we throw them into the region of the subconscious. Complex grows more complex as knowledge advances, but complex is complex, not simple, in the very first manifestation of knowledge. Evolution, then, may be applied to mind as well as to matter in the sense of growing complexity; but what shall we make of the statement that there is an *integration of matter and concomitant dissipation of motion, during which the matter passes from an indefinite incoherent homogeneity to a definite coherent heterogeneity, and during which the retained motion undergoes a parallel transformation?* Thought cannot be stated in terms of matter and motion; there is a gulf between the two. No doubt brain may grow more and more complex as mind advances; but that is a physiological truth, not a psychological; and Spencer vindicates psychology against Comte's criticisms as a separate science. Well, then, even if this science exemplifies the evolutionary tendency to complexity, it does not, and cannot, fulfil Spencer's formulated law of evolution. The case is no less clear as regards sociology or ethics. But what is the use of a law that does not fit the facts? What is the use of claiming to give an interpretation "in terms of matter and motion" when the terms themselves rebel against the office to which they are put?

Evolution, however, is not the only great interpretative category which Mr. Spencer has in view. It is flanked by two others—dissolution and equilibration. Dissolution is the opposite of evolution. Equilibration

stands between the two—the last stage in evolutionary·
process within any finite aggregate before the forces of
dissolution break in from the outside. At first sight
nothing can seem more trivial or truistic than this
threefold view of nature. Everywhere things are either
growing more complex, or else getting less complex, or
else standing still without either gain or loss. No
doubt, but pray what else could things do? Did it
need a great philosopher, controlling all the thought of
the past and all the science of the present; did it need
a system of philosophy in a dozen volumes to teach us
this pedantic formula?

Yet perhaps there is rather more underneath the
surface, whether well founded or ill.

First, as to dissolution. Dissolution is by no means
of equal importance, in Spencer's systematising of knowledge, with evolution. At times, theoretically, he may
co-ordinate the two; but nine-tenths of his energy is
spent in showing how nature weaves her web; barely
one-tenth is allotted to the process of unpicking the
fabric and resolving it again into its threads. In one
form dissolution has a place in the system of nature as
we know it, viz. in the law of death, which is so general
in the organic world. But surely it needs no argument
to prove that dissolution, taken in this sense, does not
counterbalance evolution, or even neutralise it *pro tanto*.
Death is an element in the evolving system of organic
life. Darwin has taught us to regard death as the great
implement by which progress is secured through·the
weeding out of the less fit and vigorous forms. Weismann has conjectured that the habit of dying a natural
death, however originated, may have been a direct
advantage to the mortal species, clothed as a species
with perpetual youth, in contrast with rudimentary or

hypothetical species of living creatures which were potentially immortal.[1] But, apart from such questions, we know that death is accompanied by reproduction, and is balanced by it, and that the great evolutionary differentiation of plants and animals from the one-celled type has gone on in the midst of death. Surely, then, dissolution is a mere incident or episode in evolution so far as we are to identify dissolution with death.

There is, however, a further sense in which dissolution may be regarded as the opposite of evolution—if it come as a great cosmic catastrophe, bringing to an end (*e.g.*) the adjustment which has kept the solar system in equilibrium during untold ages. Of course such a crash on such a scale must tell not merely upon planetary evolution, but upon any organic or superorganic evolution, of which the planets in question had been the scene. From this point of view any disastrous tempest, or earthquake, or volcanic eruption may be regarded as a sample of dissolution. The larger occurrence of similar forms of dissolution Mr. Spencer seems to keep in reserve in order to account for the end of all things phenomenal. Considering the various applications of the term, may we not say that dissolution differs from evolution, not merely in tendency or direction, but also in rate of speed? That the one is slow and gradual, the other abrupt and cataclysmic? This is a fresh reason for declining to admit that the two terms are of equal importance in Mr. Spencer's thinking.

Passing next to speak of balance or equilibrium, we notice that, in Mr. Spencer's system, balance is not mainly contemplated as a phenomenon of experience,

---

[1] Weismann does not admit that he thinks of a literal struggle between essentially mortal and potentially immortal forms. What then does he mean,—he, a hyper-Darwinian?

occurring in a relative sense, or up to a limited extent, and accompanying the processes of evolution. Mr. Spencer, of course, is fully aware that life, *e.g.*, is a "moving equilibrium." But beyond that truth of experience there presses on his mind a supposed truth of theory, a doctrine of equilibrium, in which balance is strongly contrasted with evolutionary process as the limit of evolution, and the goal to which it tends.

Accordingly Mr. Spencer gives us this curious picture of the eternal and necessary nature of things : every system of matter and motion, which admits of being studied by itself, and which is subject to no influences from without except such minute ones as may fairly be disregarded,—if it is in a state of comparative simplicity, must, by eternal necessity, grow more and more complex, till at length it has perfectly worked out the inner scheme of possibility prescribed to it by its original deposit of matter and motion. When it has done this evolution must cease, equilibrium superseding it. In this sense of the term equilibrium now begins to reign. And the reign now begun, so far as appears, might, for good or for evil, be eternal, so perfect will the inner equilibrium have become,—if only there were not other systems of matter outside the balanced system of which we are speaking—other systems which, sooner or later, will interfere in its affairs with a crash of dissolution. Then comes the third and shortest act in this drama. Hitherto subordinate, counterbalanced, overruled, dissolution will now be master of all ; the web of changes, so slowly woven, so long preserved, will be rapidly torn into shreds ; the wheel will have come full circle, and nature will begin once again "at the very beginning."

By this time the evolutionary doctrine of Mr.

Spencer has ceased to bear any resemblance to a truism. Vague as are its terms, they are sufficiently startling. Fichte seemed a bold man when he announced a test for all *possible* revelations; Spencer is not less bold when he prints a programme for all *possible* universes! And all this is in the name of science—the old and sober science of mathematics. Spencer assumes a definitely limited stock of matter, a definitely limited stock of force, or, as he prefers to say, of motion; and he alleges that every universe, constituted of these materials, must continuously become more and more complex, until it reaches a balance and ultimately is wrecked by an impulse from without. If this is a scientific certainty, so be it. Yet, without attempting to control Mr. Spencer's use of science, one may express surprise at two or three features in the scheme. First, there is the perplexing doctrine of the instability of the homogeneous. It would have been so much simpler for nature to remain what it was than to work out a position of balance by more than æonian evolution, only to return once again to homogeneity and instability. So far, the doctrine seems to be this: evolution is necessarily originated because of the very nature of matter and force. Secondly, one may express surprise that the forces from without should be assumed to act only at the very beginning of all things, or at the very end of all things. If they can tear up a worn-out universe, are they not likely to tear up the majority of universes before they have so much as half run their course? Their interference may be orderly enough; it may only result in a richer capitalising of the business; but assuredly if such things happen, evolution will need to start *de novo*. Thirdly, the grounds for the theory of equilibrium are not manifest to the plain reader. If

matter and force can and must initiate a process of
growing complexity, and push it on for ages, are we
sure there is a reason in the nature of things compelling
this oscillation to cease? Does not the doctrine of final
balance point to a different conception of evolution, as
if it depended, not on the healthy nature of matter and
force, but on a certain disturbing element, and as if,
when the disturbance was once adjusted, progress
ceased? So long as the stoppage is supposed to affect
only one limited evolving system, interference may come
from other limited systems outside, and renewed evolution may take place. But we must not *always* study
nature piecemeal. And, if the whole of nature works
into a final balance, which, as Mr. Spencer says, may
very well turn out to be a thing kindred to death rather
than to life, then the whole of nature will remain there
as still as a stone—the clock having run down, will continue at rest till the end of eternity.

There is, however, another point still to notice in
characterising Spencer's view of evolution. He not only
asserts evolution, as the good and grand side of nature,
in æons of necessary and continuous growth in complexity; he assumes under evolution things much more
wonderful than any complexity—he assumes life and
thought. As far as his formula goes, the universe might
run its course and reach the end of its tether without ever
quitting the region of the inorganic. That is the result
of stating evolution " in terms of matter and motion ";
your definition does not apply to the higher manifestations of nature. Our universe, however—or let us say
our world—has reached such higher manifestations. It
has travelled all the way from the assumed solar nebula,
not merely to planets, not merely to rocks, and water,
and atmosphere, but to plants, and brutes, and men,

and societies, and ethical systems, and schools of philosophy. All these are accordingly claimed and tabulated
among the workings of evolution. But the formula does
not point to them. It must therefore be improved in
some way. We may turn here to theism, using it as of
old in supplement to the formulas of science. God works
on nature from outside. Evolution causes nothing. It
may be God's method. He causes all these great results.
Or else the formula must be amended, and we must interpret the process by its highest stages, not by its lowest
—by life and thought rather than by matter and force.
This issue must really be fairly faced. Either life and
thought are an anomalous by-product (whatever that
may mean) in the story of a universe which is purely
and essentially material; or life and thought are the
interpretation of nature—the end for which it exists—
the hinted justification of its age-long travail and agony.
The two opposing views come out very clearly in Mr.
Fiske's version of Spencer's positions, and one is glad to
know that, of later years, in Mr. Fiske's case, the higher
and nobler view has gained much ground at the expense
of the other. To merge these new orders of existence
under the vague heading of " growing complexity "—to
assimilate them to purely mechanical redistributions—
is not fair-play. The result is this: in his general
philosophical appeal, Spencer assumes that all existence
reveals a gradual ascent upwards—upwards, *i.e.*, to life
and thought. And the knowledge that life and thought
have emerged on this earth inclines men to regard
favourably the claim of evolutionism to serve as a
philosophy. But, when he comes to state his system in
detail, the very attempt to trace unity of process is
abandoned. Instead of that, we have a number of
parallel developments; material simplicity (homogeneous

matter) passing into material complexity (universes); biological simplicity (the cell) passing into biological complexity (the multicellular organism); psychological simplicity (the presentation or impression or psychical " shock ") passing into psychological complexity (mind); sociological simplicity (the tribe of kins-folk) passing into sociological complexity (through militarism to industrialism, the final non-coercive order). From the formula of "growing complexity" no one could have deduced, or can deduce, organisation, consciousness, history. Again, take Mr. Spencer's sub-divisions in any one of the higher sciences. It is well to review the historical phenomena of human society under the heads of domestic, political, ceremonial, and ecclesiastical institutions. These headings are drawn from knowledge of the special facts to be dealt with. Can any one say that the abstract formula of *growing complexity* suggests these subdivisions? Is any light thrown upon them by speaking of "aggregations of matter" or "parallel redistributions of contained motion"? The great German idealistic philosophies may claim our faith, or they may find us no better than doubting Thomases, but at least we owe them this admission,—they have tried to exhibit the world we know as the necessary realisation of one great prin-ciple in stage after stage. Mr. Spencer has not been bold enough or rash enough to attempt this. But, without doing it, he claims all the advantage of having done it, and of having crowned his efforts with success. If we are able to distinguish words from things, we shall refuse to admit that so great a distinction can be so cheaply earned.

CHAPTER IX

MR. SPENCER'S THREE DOCTRINES OF HUMAN WELFARE

Goodness is *more evolved* conduct, *i.e.* is "wisdom"—An appeal to (cosmic) history !—It is *balance*, of egoism and altruism—An appeal to economics and to (hedonistic) psychology—It is *individual freedom*—An appeal to rights, and to (human) history, emerging from militarism—For which Spencer feels an exaggerated dread—Spencer masses facts rather than unifies knowledge—The "social organism" is only a phrase with him.

HAVING sought to differentiate Spencer's position as an evolutionist from Darwin's, we may now return to our more proper theme, by asking what doctrine or doctrines of human welfare Mr. Spencer furnishes.

We note three main positions, independent of each other. First, human conduct is good or wise in proportion as it is more evolved ; secondly, in proportion as it draws near the ideal goal of ethical progress, the perfect balance between egoistic and altruistic impulses ; thirdly, in proportion as it is faithful to the high attainments of modern social advance with its ideal of a still higher future, when the compulsory co-operation distinctive of militarism shall have entirely given place to the free co-operation distinctive of industrialism.

The first of these positions is not specially formulated or emphasised by Spencer, but represents an assumption that runs through much of his system, and that works to the surface at many isolated points. Good conduct is more evolved than bad conduct, and, being more evolved, it is more complex. The bad man is like a clumsy juggler who can barely keep in motion two balls at once; the good man is like a clever juggler who, without sign of effort, can control his half-dozen balls or more. With this is associated the conception of evil and in particular of crime, as atavism. The criminal is a survival or revival of a lower social type; he cannot bear the stress of civilisation at its present pitch, and so falls back upon "good old rules" and "simple plans." A further implication is plain. So far as this mode of conceiving things is true, moral progress runs parallel with intellectual progress and rests upon it. The criminal breaks down because he is psychologically incompetent. Goodness is wisdom. Perhaps such a position is a wholesome corrective of dangers that beset ordinary ethical thinking. When we have begun by distinguishing between intellectual and moral advance and by insisting that one may be found in separation from the other, we are too apt to let the distinction harden into an absolute contrast. It is well to have our attention recalled from simplicity, as a moral ideal, to the rival claims of wisdom. For ultimately all ideals must converge; and no sort of goodness can long commend itself which fails to make room for the higher tasks of culture and the finer growths of intellect. If we ask next what is the authority for this view of things as assumed by Spencer? If Comte may be regarded as appealing to biology, to history, and to a half-psychological half-ethical doctrine of altruism, to what does Spencer here appeal? We must answer that he appeals to the whole cosmic process. It is a kind of appeal to history, but to history generalised and expanded far beyond the range of the human race. From the unstable homogeneity of the hypothetical nebulous cloud, beyond

which thought can discover no deeper foundations in
the abyss of the past, thence on to the present, all
things, as they have evolved, have grown ever more
and more complex; let us too join the onward march;
let our minds expand and ramify and interweave their
forces; let us grow ever better and better by growing
ever more and more elaborate and intricate in our
behaviour! An impressive appeal, if you have any sort
of religious faith, theistic or even pantheistic. If "all
things are working together for good," then the
behaviour of "all things" may well furnish a type for
our own conduct. But, apart from the assumptions of
religious faith, it hardly seems possible that so abstract
a formula as "growing complexity" should command
the reverence of the human conscience. And one is
driven to ask whether conscience has not its own tests?
And whether Spencer's appeal does not carry its own
limited cogency just upon this account, because it has
been examined, and, in a sense, countersigned by
conscience? Whatever may be the philosophy of
conscience, the voice of conscience does not wait for
authority from evolutionary doctrine or from any other
outside critic, before telling us, and that in no faltering
tones, that goodness is wise, that sin is foolish, and
that wisdom, which is one name for goodness, demands
from us progress, both intellectual and moral.

The second of Spencer's ruling moral conceptions is
that of a balance between egoism and altruism. This
balance is twofold; there is to be a balance between
egoistic and altruistic promptings in the individual; and
there is to be a balance between personal gratification
and social service in experience. But the two processes
are to be developed harmoniously, and are to achieve
their tasks together. On one side, this draws from

Spencer's general evolutionary philosophy. It corresponds to that doctrine of final balance which is so dubious and so characteristic an element in his deductive processes. Historically, it probably owes its suggestion to the doctrine of the *stationary state* formulated by the Political Economists. To them progress meant largely numerical growth in population. When that tremendous pressure should have to cease for lack of further space, they looked forward to a stationary state of society ; and J. S. Mill at least plucked up courage to regard the stationary state as a thing to be desired rather than dreaded. In Spencer's system, this conception is given the lordship over ethical thought, strictly so-called ; and complexity, or the progressive ideal, is overborne by the ideal of balance, or fixity, as a Utopian or millennial vision. Has this ideal any further authority beyond the place allotted to equilibrium in Spencer's First Principles ? Assuredly it has. It represents the hedonistic postulate. It represents an appeal to consciousness, and to that form of consciousness which declares pleasure to be the end of life. Distracted between the craving for personal pleasure and the momentous claims of others, the individual is bidden take comfort from the evolutionary process, which, moderating personal claims, and increasing altruistic efforts, is preparing a heaven upon earth for the benefit of our very remote posterity ; at least, if the world lasts long enough. But the fundamental postulate remains ; pleasure is the good. All systems, we are told, virtually involve this assumption, and all moral truths are lighted up by it. Why is altruism good ? Because it gives pleasure to other persons, although at personal cost. Why is egoism good ? Because a judicious tincture of egoism increases average happiness. Thus, in this department of the system, the supreme law is not "Be complex," but "Get pleasure," or, in its noblest form, "Give pleasure," but in the form which best reproduces the meaning of the doctrine of balance, "Promote maximum pleasure." This psychological test of the good overrides and controls all the other tests with which it is associated in the *Data of Ethics*—physical, biological, sociological. Spencer himself bears witness to this fact—to the supremacy in his thinking of a psychological test ; nor have any reason to challenge or complain of it. By all means let the moral consciousness speak ; and let it be a supreme, if not a solitary, guide ; but are we sure that this hedonistic doctrine is the authentic and final utterance of the moral consciousness? Is complexity—which in Spencer's thinking stands for moral and intellectual progress—really to yield its place of supremacy to compromise or balance, if the latter secures maximum pleasure all round ?

The third ideal dominates Spencer's formulated sociological doctrine. Here he is the out-and-out champion of individualism. His sociological lawgiving distils down into a single old phrase, *laissez faire*. Of course so acute and systematic a thinker betrays the same bias in his ethical writings as in his sociology. He is a thorough individualist in his emphasis upon justice, with its indefinite appendixes in favour of negative and positive beneficence. Both as moralist and as sociologist, Spencer is full of the thought of individual rights : in curious contrast with previous utilitarian writers, and in curious sympathy with intuitionalism. This doctrine of rights constitutes, in fact, one of the most genuine and most important among the vanishing traces of intuitionalism in Spencer's thinking. Still it seems fair to say that when he handles ethics technically this doctrine of rights is overruled and held in check by a doctrine of maximum pleasure. The Utopian state is not praised on account of its freedom, so much as on account of its balance and harmony. All this is altered when we pass to the technically sociological discussion. Here freedom is the good ; not harmony or co-operation *per se*, but that harmony or co-operation which results from freedom in contrast with that which results from compulsion. This (sociological) doctrine is supported by an appeal to history. The cosmic philosophy is silent here, except in so far as it hints that the voluntary co-operation of industrialism, being later in origin than militarism, is presumably higher—more truly evolved—more complex. There is hardly any trace of hedonism in the argument. If the appeal to history ran into the form, "Freedom has worked better"; "Freedom has increased average happiness"; that would, of course, be sound hedonistic doctrine. But Spencer, like Comte, has little taste for detailed historical parallels as a means of appeal to history ; both prefer to look to the mighty onward current,—while unfortunately their witnesses, reporting what they see there, agree not together. Comte regards individual freedom as a sign of the weakness inherent in " critical periods," which can be nothing better than narrow bridges leading from one organic period to another ; Spencer regards individual freedom as the highest stage in evolution—the great good towards which past conditions have steadily moved on. Comte, in the name of fact and science, preaches a new synthesis ; Spencer, speaking in the name of the same great authorities, pronounces a curse upon it. Every attempt at closer social organization seems to him a relapse into outgrown military forms of society, and an act of treason towards industrialism.   He does not discuss this, but takes the assumption for granted, with an *a priori* vehemence that we should find it hard to match, *outside* the ranks of scientific empiricists.   Of course he has informed himself, as few men have done, of the vast prevalence of militarism during former ages.   Where society has been highly centralised or organised, it has been in the past, one might almost say, uniformly, a society of a military type.   And a very little study of sociology will make it plain that, if a society is drilled and regimented and over-governed, it will lend itself much more readily to manipulation for military ends than a freer or more individualist society would do.   Still, all this hardly constitutes a proof.   It may be unfair to style it a prejudice : let us call it a presumption, and a grave presumption ; but is it a proof?   The Hindu who mocked at the very idea of ice had a wide experience of the fluidity of water ; and it is perfectly true that H_2O tends strongly to the liquid state, being a liquid " at ordinary temperatures and pressures " ; yet *solid water* is a fact of some importance to Arctic and Atlantic voyagers, whom it brings into danger ; not to mention British outdoor labourers whom the frost robs of work, or plumbers to whom it is better than a mine of gold.   Ice then is a fact, though to some it may be a novel fact.   And socialism might be a practicable policy even though it be a new development of strict social organisation.   It is not disproved by calling it bad names.   Neither socialism itself, nor the modern political changes stigmatised by their opponents as socialistic, are in the least degree animated by any conscious breath of the military spirit.   They do not mean to serve it ; and, whether they turn out good or evil, we cannot be sure that they will turn out to

be in the line of militarism. There is no promise or
potency of a *coup d'état* in the Government purchase
of telegraphs or even of railways. When Mr. Spencer
insists upon treating every civil servant as a disguised
soldier and secret conspirator, he does not carry our
convictions with him ; he only proves to us that the
new science is very like the old obscurantism, and that
you may find a perfect sample of the High Priori temper
in a mind wedded to familiar facts, and inaccessible to
unfamiliar ones.

Mr. Spencer then has given us three ideals ; and they
hardly seem to agree with each other. One is an ideal
of progress, two of fixity ; one praises complexity,
another tells us that the best government is the
minimum of government, but that means simplicity, not
complexity. It is the nature of reason to invent short-
cuts and to retrench needless labour. The most
advanced is not necessarily the most elaborately
organised ; it is not so, if Mr. Spencer is right, in
society. Moreover, the sources of authority are
different. One appeals to the cosmic process ; one to
the experience and tendency of human history ; and
one direct to consciousness. In Martineau's language,
Spencer's ethics, technically so-called, are "psychological
ethics" though "heteropsychological." Surely we have
reason to fear that the promised unification of knowledge is still sadly to seek. Vast masses of knowledge
have been collected. They fairly bristle with suggestions—highly interesting, extremely divergent suggestions ; but neither within the four corners of Mr.
Spencer's own system, nor when we bring his teaching
into comparison with that of other votaries of fact, do
we find science stilling the metaphysical strife, or giving
clear guidance in human things.

One part of Mr. Spencer's teaching, held by him like some others in common with Comte, has not yet been referred to ; his doctrine of the analogy between society and an animal organism. I have omitted this, because I regard it as an ornamental excrescence on Spencer's teaching, not as an essential or even a significant part. Whatever function the appeal to biology played in Comte, it seems to play very little part in Spencer. " The social organism " is an outplayed authority—a god *emeritus*—a depotentiated deity—on Mr. Spencer's pages. " The social organism " is a metaphor with him and only a metaphor. The individual cells are asserting themselves, and the unity of the organism is coming off second best. If Comte tells us, " Be parts ; be mere parts, living for the sake of the whole," Spencer thinks such advice the very worst possible. Each for himself; fair-play all round ; justice the supreme consideration, politically and socially ; the occasional surrender of individual rights purely a personal matter, with which public action and public opinion dare not interfere—such is Mr. Spencer's social programme. It is the antithesis of Comte's. Where Comte says "Yes," Spencer says " No," very nearly all the way through. We take it, therefore, that, beyond serving to explain his views lucidly and add a grace to them, the doctrine of the social organism does nothing for Mr. Spencer.

CHAPTER X

MR. LESLIE STEPHEN'S "SCIENCE OF ETHICS"

Stephen a utilitarian—Who came to believe in evolution as a scientific fact—
Begins here with facts : ethical judgments exist—Organisms seek maximum
efficiency—If social "tissue" is "organic"—Then ethical laws may be the
conditions of maximum social efficiency—(Nature cares for individuals)—
Nature says, "Be strong !"—Ethics says, "Society, be strong !"—The
ethical is the *typical* society, and *therefore* ethical judgments are binding—
But the type is actual, not ideal !—Society is a complex whole, changing
while its parts are unchanged—Criticism—Sanction for individual goodness
lies in sympathy merely—Sometimes we are too good for our own interests !
Compared with Comte, lacks *authority*—With Spencer, calls "health" the
ideal, and ridicules "balance"—With Darwin, not struggle of individual
with individual, but of individual with society—With Utilitarianism ;
discourages the calculation of consequences—Most of his positions may be
accepted in a deeper sense.

MR. STEPHEN makes his intellectual history very plain
in the preface to the *Science of Ethics*. He started in
the life of thought as a utilitarian, under the strong
influence of J. S. Mill ; and he never came to regard
the utilitarian position as discredited. But, in course
of time, impressed partly by Darwin's theory, partly
by Spencer's writings, he began to crave a restatement
of ethics. This was in no sense a concession to intuitionalism. Spencer's "reconciliation of intuitionalism
with empiricism" is indeed accepted by Mr. Stephen,
as appears from his other writings ; but, unless one has
read the *Science of Ethics* very carelessly, no reference
is made to the doctrine in Stephen's moral system, and
it seems to go for little with him. Indeed, his first

view of evolutionism hailed it as a new stick for beating
the intuitionalist dog withal—a new reason for rejecting
the conception of ready-made and all authoritative ideas
in the human mind.   And when he conceived the
possibility and desirableness of a new system of morals,
he had not in view a worthier ethic than utilitarianism,
but rather one more fully in harmony with new scientific
truths.   Science, not philosophy, demanded the change.
Evolutionism must be given effect to.   If the change
results in a more adequate statement of moral ideas,
that is, for Mr. Stephen, a secondary matter.   The great
thing with him, as on a broader canvas with Mr.
Spencer, is to unify thought.   One fresh province is
to be gained for the master principle, evolution.   As
Prussia Prussianises its Polish dominions, as Russia
desires to Russianise Finland, so Mr. Stephen evolu-
tionises his ethics.   Of course in each case the conquered
is assured that ultimately his own interests will be
served through accepting the *régime* dictated by the
conqueror.

When dealing with Comte, we suggested a difficulty
for thorough-going phenomenalism in the very conception of duty ; and we argued that Comte uses the
doctrine of the social organism as justifying the claim
for individual submission to the public weal.   Mr.
Stephen also makes an appeal to biology, but he does
not directly employ that appeal as a basis of ethics.
He begins more simply, by accepting current moral
judgments.   Science deals with facts ; well ! these are
facts.   Ideal ethics, indeed, are no facts of everyday
experience ; but Mr. Stephen tells us that he has
nothing—so he says ; nothing—to do with ideal ethics.
It is the current rules, which have been historically
recognised and appealed to, for which he desires to find

a scientific basis. Mainly he is concerned with defining ethics—with reaching greater accuracy than is possible for the colloquial judgments of mankind. His voyage is one of survey and measurement. Ultimately his reasonings must bear on the question of the justification of ethical judgments; primarily, he is concerned with their precise statement. And, indeed, precision is one great mark of science, along with exhaustiveness and coherence.

What, then, has evolutionism done for him? First, it has taught him that every organism strives to attain to its maximum efficiency. Darwin, indeed, has pointed out that the organism which fails to strive, or fails to attain, fails also to survive. There is, however, little direct Darwinism in the *Science of Ethics*;[1] and in its absence Mr. Stephen's view of an organism sounds almost Lamarckian—dreadful word!—or even—more dreadful still—Spinozistic. He has borrowed from science the fact that each organism seeks maximum efficiency. Darwin's view of the reason of that fact he accepts rejoicingly; but he does not utilise it.

Secondly: he agrees with many predecessors in holding that society is essentially organic; and he gives the usual and correct interpretation of that statement, viz. that in society, as in plants or animals, the whole explains the parts or is prior to the parts; that you cannot explain the whole as a mechanical combination of separate parts, but on the contrary, must have a knowledge of the whole before you can correctly define or explain any one part.[2] Since man is essentially dependent on society—since man is by nature

---

[1] Some passages on pp. 72, 73, 91, 92, where Mr. Stephen does Darwinise, are quoted in Williams's *Evolutional Ethics*, 419, 420.

[2] p. 32 ; cf. also p. 110.

social—therefore we call society an organism. It is doubtful whether we can credit this thesis to the contributions which Mr. Stephen has received from evolutionism. It goes back—not to search more deeply —as far as Comte, who had no patience with idle inquiries into the origin of species. But in Mr. Stephen's mind it is lighted up and vivified by modern evolutionary science—especially by the doctrine of a "moving equilibrium" between organism and environment.

In the next place, Mr. Stephen may be said to combine these two positions in a syllogism, which issues in a third proposition by way of conclusion. Since all organisms strive after maximum efficiency, and since society is an organism,[1] society also will strive for maximum efficiency. But—here to a certain extent hypothesis begins—we may very well understand moral rules as the outcome of this striving, or as the formulated conditions of maximum social efficiency. The effort or nisus of the social organism has broken into consciousness in the individual members of society in the shape of moral commands or ideals of duty. A Darwinian doctrine of competing organisms is scarcely if at all found in Mr. Stephen. So far as he thinks of any competition, the competition is rather between the claims of the individual man and the claims of society. Each man is an organism, immersed in the thickest of the struggle for existence, striving to do the best for himself. But then, society too is an organism; and it also strives; and its precepts cut across the blind self-interest of the natural man—checking it, modifying it, perhaps overruling it.

---

[1] We shall see, however, presently that Mr. Stephen prefers a slightly different phraseology.

Morality then—it is a hypothesis, but a strong one—consists in the recognised and approved conditions of social efficiency. There are, however, some qualifications. So far as social well-being implies individual physical well-being, we do not (unless in a secondary degree) count the observance of such conditions among moral duties. It is not a moral act to eat when one is hungry—it is natural. Nature secures our doing that ; society need not trouble about the matter; and morality —which is the voice of society, protecting the interests of the race—if it speaks of prudential regard to one's health and interests as a duty, gives prudence a comparatively low position among the virtues. Whatever is the outcome of organic natural impulse forms rather a presupposition than a part of morality. Further— consulting, as I understand him, the usage of language— Mr. Stephen is inclined to confine the epithet " moral " to altruistic actions. Ordinary conscious action in one's own interest seems independent of the moral spur. It seems to stand almost, though not quite, on the same level with natural instinct. But with these two qualifications—that morality does not include those conditions of social efficiency which are taken care of by instinct, nor yet those in which the social demand coincides exactly with the promptings of rational self-interest—Mr. Stephen holds that morality means the law of the social weal, or the conditions of maximum social efficiency. The law of nature is summed up in one terse injunction : " Be strong ! " The law of morality is similar : " Let society be strong ! " And social strength or welfare is found to lie in the individual virtues of courage, temperance, and truthfulness, along with the more directly social or altruistic virtue which is sometimes hailed as " justice," and again as "benevolence," but which, in every case, takes as its direct and supreme rule the highest interests of society, or the welfare of other persons.

Mr. Stephen explains this conception of morality by the aid of the idea of type. A type in each class, apart from extrinsic and accidental tests, is that which attains maximum efficiency. The most moral human society is the most efficient or most prosperous human society. Here then Mr. Stephen has found a second answer to the question, How can empiricism speak of morally better and morally worse? The first answer was provisional; the moral consciousness is a fact, and we accept its utterances as approximately trustworthy. The second answer goes deeper. Morality is not something externally added to social life, as a necklace or a posy of flowers may form a slight addition to the graceful dress of a beautiful woman. Morality is simply the perfect performance of social functions, like the glow of health upon a beautiful countenance. Therefore human life in society points to perfect morality as its own typical perfection in the way of *vitality* or of *health*. And here we see what biological evolutionism has done for Mr. Stephen. It is not indeed strictly necessary for his argument. There might be evolution in human society, with the moral law as its ideal goal, even if there were no evolution of species in the infra-human world. The "typical bow" which is "felt out" might point us to Mr. Stephen's conception of morality as the true type of our own social being, even if there were no evidence that "the animal . . . feels itself out."[1] But there would not be the same trace or hint of authority in Mr. Stephen's evolutionary interpretation of morals, did we not believe in the origin of species by a process of

[1] p. 79.

evolution.   Morality is vindicated when we see that all
nature, or all animated nature, toils upwards, and that
our goal, if not as individuals, yet as a race, is moral
goodness.   The morally good society is the typically
human society ; the morally good individual, so far as
he is good, is qualified for membership in that society.
Here, however, a difficulty arises.   Mr. Stephen renews
his warning against a doctrine of absolute or ideal ethics.
The type is a real type in the actual present, a type
constantly modifying itself as the environment alters or
as the conditions of struggle change.   Yet on the whole
the broad outlines of the type are fixed ; the cardinal
virtues are recognised on all hands, very nearly as they
have been blocked out by Mr. Stephen ; and we may
say in general terms that morality represents the human
ideal—the demand addressed by the race to every individual.   Here as elsewhere, Professor Alexander
gives us a more extreme position on the lines of Mr.
Stephen's tentative suggestions.

It is necessary to emphasise one other feature in Mr.
Stephen's evolutionary view of ethics.   He insists that,
in such a society as that of mankind, the organic whole
may change while the individual organisms are unchanged.   In a somewhat obscure passage he contrasts
this most complex case, exemplified in human society,
with simpler cases, in which the individual organism
and the social organism are modified simultaneously.
One cannot help thinking that the whole distinction is a
piece of very doubtful philosophy.   What Mr. Stephen
wishes to bring out by it is the fact that the social
organism exerts its influences by the spiritual forces of
thought and language, apart from any necessary
reference to physiological change.   So completely is Mr.
Stephen indifferent to the moral applications of Mr.

Spencer's view—which he shares [1]—as to the origination
of apparently intuitive perceptions.   Morality is evolved,
according to Mr. Stephen's statement, not at all by
means of a growing stock of innate moral sentiments,
though he believes in these, but essentially by a super-
organic process in the region of human culture and
intercourse.   Training makes the man.   Physiologically
there is as good as no difference between the civilised
and the savage.   This is proved by the fact that the
infant child of civilised parents, if stolen by savages,
will grow up in the likeness of the savage race, and that
the child of savages, if reared among the influences of
civilisation, will make a very fair average citizen.
Differences there may be, which will hold their ground,
even when transplanting has occurred and the new
environment has done its work ; but these (or so I
understand Mr. Stephen) are insignificant in comparison
with the broad fact, that every child or man is a human
being, *homo sapiens*, and therefore a moral being ; that
each child or man is merged in the community where he
has grown up and takes on its colour.   Now one is fully
prepared to agree with the positions here laid down.   A
man's a man for a' that ; there is a vast moral unity in
the human race.   But Mr. Stephen's mode of stating
his position seems highly dubious.   Anthropologically
or physiologically, man may be simply man, neither
more nor less ; but we were speaking of sociology, were
we not ?   If the social organism is changed, are not the
constituent individuals changed, sociologically?   Strange
metaphysical subtlety of empiricists, if this is to be
denied !   To remind us that the members of society are
physiologically unchanged is beyond the mark.   To
point out that civilised citizens would have been savages,

[1] *English Thought in 18th Century*, i. p. 56.

if reared among savages, is again beside the mark.   The question is not what they might have been, but what they are.   Mr. Stephen may settle it with other authorities whether or not it is true that the "innate faculties of a modern European differ little from those of the savages who roamed the woods in prehistoric days."[1] Be that as it may, the educated faculties of a modern European differ greatly from those of a contemporary or prehistoric savage after his fullest savage training. Else the two societies could not differ.   Mr. Stephen thinks he is offering us a contrast between the individual human organism and the social organism ; he is only marking the contrast between two distinct sciences, sociology and biology.

So far, then, we have got the following account of ethics from Mr. Stephen ; it is the law of the social weal imposed, essentially by precept and example, upon individuals.   But there still opens before Mr. Stephen another problem.   How does the individual come to receive and obey the aforesaid law ?   And why should he do so ?   He is led to care for others—so we may put Mr. Stephen's view—by sympathy.   To be aware of pain—of another's pain—is to be more or less pained oneself ; to be aware of pleasure—another's pleasure— is to have a pleasing object of contemplation, and thus to be oneself more or less pleased.   Two harps stand near each other, you strike a chord upon one, the other takes up the sound—that is a picture of the origin of moral feeling as Mr. Stephen states it.   If any one is inaccessible to these secondary emotions, evoked by primary emotion on the part of his fellows, his intellect is at fault ; he cannot have clearly understood that they are really suffering or really happy.   It follows that he

[1] p. 102.

is an "idiot," says Mr. Stephen. Now, sympathy is a
vague and ambiguous word. If you say that morality
rests upon sympathy you may mean almost everything
that the moralist can require, or you may mean hardly
anything at all. Mr. Stephen, like Adam Smith I take
it, means very little indeed. Morality rests upon a
rooted psychological incapacity for clearly distinguishing between *meum* and *tuum*. It would seem perfectly
open to the selfish man to retort the charge of idiocy
against moralists of Mr. Stephen's type. "Idiot yourself," the bad man might say, with great force. For
indeed there is nothing so incommunicable and purely
personal as mere pleasure or mere pain. And moral
sympathy, which makes us partners with one another
in all things, is very far removed from automatic promptings or illusions as to the limits of personality; it does
not fall below clear thought, but includes it and goes
beyond it. Love is a relation of person to person, and
the keen pang of love is not due to any vague appre-
hension, "What! there is suffering about, is there?"
but to the dreadful consciousness, "He is in pain!
Precisely he! Not I, but he! That is the maddening
thought!" Yes, and there too lies the ennobling
experience.

The further question, "Why should I yield? why
care for others?" receives the answer, "Generally in
the long-run it pays in pleasure to oneself to do so;
but sometimes, we must admit—in unfortunate cases,
or where there is too lavish generosity—self-sacrifice
means a heavy nett loss." And with that the *science* of
ethics, as conceived and worked out by Mr. Stephen,
confesses itself bankrupt. The point has come at which
the question of the justification of the moral judgment
can no longer be thrust aside. Defined at first as social

requirement, duty is now tested from the point of view of the individual consciousness; when a gulf discloses itself between the individual life and the social whole. We live in an irrational world; for our nature craves and postulates happiness; and, although sometimes when we deserve it we get it, yet often we have to do without. Better look facts in the face! There is no more to be said. The timid man will obey morality as a sort of insurance policy, he will be moral on the chance that immorality may be punished. But often the bold man will play a recklessly speculative game—heavy risks, great profits. If he succeeds, how can you prove to him that he chose wrongly? The "idiot" may have been quite right from his own point of view. So much for the "Science of Ethics!" The Christian, too, admits that our moral nature lays down great postulates, to which experience does not always conform. But we look to the future for the recompense of reward—not "so much pleasure for so much goodness," but a larger life, and the "wages of going on and not to die."

It will clear our thoughts if we compare Mr. Stephen with his predecessors.

First, with Comte. In some respects Mr. Stephen seems to be the legitimate heir of Comte, especially in regard to the biological appeal. Stephen's thinking is guided throughout by the biological analogy, and he is able to throw fuller light upon it by the modern evolutionary conception of infinitesimal changes which maintain a moving equilibrium. Like Comte again, and unlike Spencer, he definitely identifies morality with the claim of society upon the individual in contrast with all individual claims or wishes. But here the likeness to Comte ceases. First of all it is perhaps significant that Mr. Stephen refuses to speak of a social organism, preferring the more indefinite phrase, social tissue.   That points us to the individualism which lurks in the background of his mind,—to his impending re-assertion of the cells *versus* the organism,—to his postulate of personal pleasure as an ultimate test.   But there is a more immediate difference from Comte, in Mr. Stephen's distrust of sociology and of all forms of authority.   Keeping that in mind, we might almost say that Mr. Stephen uses the biological analogy to reach sociological but not moral truth.   With Comte sociology *was* the new ethic; or, at the lowest, socio-logy, the science of corporate action, was the necessary basis of ethics as the science of individual conduct. Mr. Stephen, however, speaks contemptuously of the attainments of sociology.   He thinks it scarcely a science, and values its standpoint merely as a stepping-stone to a new statement of ethics, in which the biolo-gical analogy defines rather than justifies the moral law. It follows that the biological appeal has not the moral or *quasi*-moral weight which it had with Comte.   Nothing takes its place.   The appeal to consequences admittedly breaks down.   In fact there is a marked absence of *authority* in ethics as presented by Stephen.   Comte says, "You are members one of another, be loyal mem-bers of the social whole."   Stephen says, "Social tissue requires you to do so-and-so, and of course you are very dependent on the social tissue; still, you have a centre of being in yourself, and there is always the possibility left that it may pay you to defy society; very rarely indeed will it do so, but sometimes, no doubt, it will, if you are unsocial enough, idiotic enough, bad enough." Comte allots no sphere at all to the individual, while Stephen, like other hedonists, gives him a sphere, but makes it fall outside of morals. What is moral is not personal, but social. What is personal is not moral, but hedonistic.

As compared with Spencer, Stephen also deals mainly with one great harmonious process of evolution, though with him it is purely biological—either the maintenance of health, or the fuller unfolding of life; and he does not trouble us with definitions in terms of matter and motion, or with hymns of praise to complexity. Spencer's second great ideal, that of balance between egoism and altruism, is dismissed by Stephen as a Utopian dream; but he would dearly like to lay hold of it, if he dared, for he is as much a hedonist as Spencer; and, in the absence of perfect righteousness even from Utopia, Mr. Stephen's whole moral world lies at the mercy of chance. On Mr. Spencer's third ideal, that of political and social *laissez faire*, Mr. Stephen finds no occasion to express an opinion in his own more purely ethical treatise.

Next, if we contrast Mr. Stephen's positions with those of Darwin, or rather with those suggested by Darwin's views, and worked out later in their ethical and social bearings by other writers, we observe an almost entire absence of any doctrine of struggle for existence. Evolution is accepted in the Darwinian sense, but little or no reference is made to the Darwinian theory of the conditions of evolution. That remains true even in regard to the few passages where Mr. Stephen in a sense Darwinises, speaking not of one human social tissue, but of diverse forms of tissue. These various tissues may be thought of as competing with each other, but are hardly recognised as struggling for life, and as either dying out or else covering the whole field. If Mr. Stephen has a struggle in view at all it is that between morality and selfishness, social tissue and personal organisms, society and individuals— a dreary conflict, to which there seems no discernible limit on the farthest horizon.

Finally, how does he differ from Utilitarianism? There is one very important practical difference. The Utilitarian, as a moralist or spiritual director, defines right and wrong, and urges men to define right and wrong, by a computation of visible results, in the light of the present tastes and faculties of living men. Mr. Stephen on the other hand, when he speaks as an expert upon moral points—as a consulting moral physician, or the giver of "counsel's opinions" in morals,—Mr. Stephen remembers what evolutionism has taught him, that the race has changed and is changing. Therefore he keeps in mind the probability that results, which we think highly advantageous, may be judged very differently by a future society when it measures them by its new standards and altered tastes. And therefore Mr. Stephen appeals to recognised moral duties and maxims as guides to social welfare. He distrusts the most acute calculation of the consequences which we can foresee. Morality has been evolved on the lines of social advance, and points us on to the true line of further progress. Not pleasure, but health or vitality, is to be our test. Now this is good and wholesome teaching, far better than hedonism, however universalistic. But with Mr. Stephen this is all a technical thing. He speaks thus as a moralist to moral minds. But, when he speaks as a man to individual men, there is instantly a relapse into hedonism, and that of the selfish sort. Granted the moral judgment—given a soul devoted to the social weal—Mr. Stephen offers vigorous and pointed encouragement, and dissuades one from being argued

out of obedience to conscience. But, if the moral judgment be disputed, and if any soul prefers his own private
weal, Mr. Stephen gives no help. To call selfish men
" idiots " merely because they distinguish *meum* from
tuum is not helpful. Tastes differ—that is the last
word on these questions, if we adopt Mr. Stephen's
premises.

One thing more we might be tempted to inquire,
—How far is this whole mode of looking at morals
true and serviceable ? But hitherto we have raised no
such issue, and it would hardly be wise to discuss it at
this particular point. Only so much we may say : if
the community is to be the authority in ethics it must
not be narrowly identified with any external society ;
and that which it lays down as duty must not be merely
what is socially convenient—still less, what is convenient
for society and costly to the individual ; duty must include absolute and ideal elements, whose fulfilment is
quite as much for the interest (in the true sense) of the
individual as for that of society. But, granted some such
deeper view of society, it may be useful to have a statement of morality as *the single or continuous human
ideal*, and to have this in terms of biology. It is well,
too, that one of the biologising moralists should emphasise, not obscure subconscious possibilities of organic
change, but the knowable influences of human education
and historic culture. We shall quote Mr. Stephen for
this at a later point.

PART III

DARWINISM, OR STRUGGLE FOR EXISTENCE

CHAPTER XI

"DARWINISM IN MORALS"—MISS COBBE'S PROTEST

Darwinism may be applied to morals by analogy—Or, as here, by explaining man's evolutionary origin—Miss Cobbe attacks Darwin's explanation of the rise of morals out of intelligence *plus* sympathy—And the hypothetical palliation of murder—Little trace of *natural selection* in Darwin's ethical statement—Darwin's analysis may be accepted, not his view of reason.

IT is not necessary again to recapitulate the leading points of Darwinism. Nor is it desirable to pause at present in order to weigh some very grave metaphysical objections[1] to the terminology and conceptions with which Mr. Darwin went to work. We are more concerned to ask how Darwinian ideas have affected the theories of morals or of society which follow biological lines.

[1] Urged with great force by Dr. Hutchison Stirling, and incidentally brought out with masterly power in Mr. George Sandeman's *Problems of Biology*. Mr. Sandeman's statements go far to convince one that Darwin's theory is only a possible way of putting the process of evolution for purposes of study, and by no means an account of the way in which the process actually took place. It *might* have happened just so, by random shots, and constant weeding, in the course of endless time. But did it ?

Possibly Mr. Sandeman himself might prefer a more sweeping verdict. See further in Chapter XVII.

Now plainly there is an ambiguity here. In the previous chapters, for the most part, we have been dealing with a scientific analogy,—consciously lifted out of one region of thought and introduced into another,—coming no doubt with a great deal of authority, but still presenting itself to view, and continuing to be regarded, as a foreign visitor. We shall still find such a course followed in some instances by writers who are employing Darwinian clues and modes of thought. The doctrine of struggle for existence may be applied to other things besides plants or animals,—to competing states, or types of society, or types of ethical thought. But there is a nearer way in which Darwinism may bear upon our problems. Man himself as an organism is brought within the range of Darwinian theories. In connection with the assertion of man's descent from brute races, fresh light—of a lurid kind, as many will think— is made to fall upon the problems of ethics ; and questions as to social origins will run back into questions regarding human origin by process of evolution.

When the world first heard of "Darwinism in Morals" from Miss Frances Power Cobbe, it was to this latter bearing of the Darwinian theories that she called attention by a resonant protest. Darwin—like Leslie Stephen after him, but with a distincter reference to animal ancestors of the human race—explained morality from sympathy, and from the interests of the species. In particular, he laid it down that the social instinct, with intelligence added to it, would sufficiently explain the origin of moral ideas. This shocked Miss Cobbe's intuitionalist prepossessions ; she could not bear to see moral ideas analysed, as if they were compounded of other, and these non-moral, elements. But above all, Miss Cobbe was aroused to natural indignation by

Darwin's suggestion, *à propos* to the action of bees in killing off drones, that, if the welfare of our species had required, under any conditions, a similar practice of murder, then the human conscience would undoubtedly have ranked murder not among vices but among virtues.

None of these positions seems to be peculiarly connected with the theory of evolution by a process of struggle for existence. They seem to belong rather to evolutionism in ethics than to Darwinism in ethics; although, as positions put forward by Darwin, they naturally and quite fairly received the title under which Miss Cobbe attacked them. Still, any thinker who believed in the continuity of life between man and beast, might, if he pleased, formulate similar positions to Darwin's. On the other hand, it is perfectly plain that such positions are incompatible with old-fashioned intuitionalism.

It is equally plain that the new fable of the bees is also (like the old one, as generally understood) incompatible with loyalty to morals. But the attempt *per se* to deduce morals from intellect *plus* social sympathy is not to be so summarily rejected. It is time to recognise that old-fashioned intuitionalism, with all its honest loyalty to the truth and its essential right-heartedness, is weak, as philosophers say, formally, and is no longer fit to sustain the "struggle for existence" against subtler theories. The whole method of building up mind from simple elements is an illusion, whether practised by intuitionalists or by naturalistic schools of moralists. There is no primitive atom in mind. Every element implies every other. If it is true in biology that the whole is prior to the parts, how much more in psychology? Moral judgments are not proved to be artificial, or secondary, or subordinate, if it is shown that they

can be interpreted in terms of man's social nature.
Man is moral because he is social : yes ; very true ; but
we are no less entitled to read the proposition from the
other end, and to affirm that man is social because he is
moral.   He is both social and moral in a higher sense
than the brute races.   We must not assume that the
earliest stages in development show us the nature of an
organism better than the later stages.   A frog is not an
effete tadpole ; on the contrary, a tadpole is an immature
frog.   And so man's moral nature is not a corollary or
appendage of brute sociability ; on the contrary again,
animal sociability is a dim and imperfect prophecy of
human morality and human society.

Of course, if Darwin's doctrine of reason were un-
impeachable, it would be idle to challenge his moral
philosophy while admitting his view of the descent
of man.   But we find his philosophical basis very
insecure.   Darwin assumes that instinct is given as a
fixed datum ; rational consciousness, when it supervenes,
works out plans and methods, but does nothing to
revise or remodel the inherited aim.   Instinct *plus*
reason form a mechanical sum in addition.   Reason is a
calculating faculty pure and simple.   Instinct remains
what it was in the brute nature (social instinct for
example, as the germ of morality) ; it now wields an
instrument of incomparably greater power, but its own
nature and its aims are unaltered.   We shall have to
give further study to this view of reason later on.   Here
we must simply affirm the counter position, that reason
transforms and revolutionises everything.   In this case
as in many others, development means transformation.
A man is not an ascidian, even if he is descended from
one.   Nor is human morality the pursuance of animal
sociality with the resources of human intellect.   No ; it

is a new aim, as well as a new method; on the theoretical side, reason; on the practical side, morality,
strictly and properly so-called. As such, it has suppressed, is suppressing, and will suppress those evil
things — evil at least between man and man, if not
between beast and beast—which instinct tolerates or
fosters.

If, however, we take this view of the meaning of
evolution, there seems no reason why the abstract
formula of " Darwinism in Morals " should be fatal to
the higher interests of mankind, or to the basis of
Christian faith.

CHAPTER XII

DARWINISM IN POLITICS—BAGEHOT

Applies Darwinism by analogy—Evolution *transforms* imperceptibly—By nerve tissue in our case ; but nothing depends on this assertion of use-inheritance by Bagehot ; it is a mere illustration—Not *ethnological*, but *political* questions—Problems both of *progress* and of *differentiation*—1st, Custom as the remedy for primitive wildness in the "fit"—Criticism—2nd, Customs winnowed by the test of war—3rd, Free discussion—Race-blending, etc., as minor factors—Three limitations on the Darwinian principle in Bagehot's application of it.

[Note B. On Professor Ritchie's *Darwinism and Politics*—Inconsistency between the different essays—One interesting hint.]

THE next important application of Darwinian notions to social questions is found in Walter Bagehot's *Physics and Politics,*—a little book full of interest on every page, and still alive with suggestions after twenty-five years. It is or seeks to be truly Darwinian, dealing, as the title-page tells us, with "inheritance" and "natural selection," and trying to "*apply* them to political society."

The author is profoundly impressed, first of all, with the *transforming* power which science attributes to evolutionary change. Things become absolutely different from what they were. Nay more ; this is true not merely of some things but of all. Everything is in motion. And therefore everything has become, in the light of modern science, "an antiquity."

Speaking more strictly of human or social evolution, Mr. Bagehot makes a very strong statement of the part presumably played by nerve tissue in rendering such evolution possible. No one, he thinks, will be able to understand evolution in history, if he has not this material basis of evolution before his eyes. In other words, we have here an act of adherence to Spencer's position—to Spencer's even more than Darwin's—against attacks such as have more recently been made by Weismann. For we have here not merely an assertion of the inheritance of acquired qualities, but an assertion of the physical inheritance of the results of mental processes. Further, we find Bagehot here emphasising an element which Leslie Stephen—though apparently believing in it—was content to drop out of sight all through his ethical treatise. Further still, we observe that for the moment Bagehot is not transferring Darwinian ideas to a new sphere, and asking how they apply there, but rather showing us how politics are influenced by Darwinism in its direct bearing upon the physical basis of mind. Man is a political animal, but he is primarily an animal. We cannot appreciate how his politics evolve unless we have formed just ideas of the process by which he himself evolves. Still, in all this, Bagehot is only preparing the way for his special contribution, which consists rather in extending the biological analogy than in claiming a wide range for biology proper. In point of fact, he might drop out this illustration altogether; he might surrender his strong belief in the inheritance of experience *via* the nervous system; and yet the main lines of his book need not be changed.

All through the discussion his problem, as he conceives it, has these two sides, physiological and political, but he declines to deal directly with the physiological questions involved. How have nations been differentiated? We assume an original unity of the human race; from whence then the differences? Bagehot is to deal with the minor causes, which are mainly political. Beyond and behind their range, other very obscure causes must have been at work to separate, not nation from nation, but race from race; to differentiate negroes or Mongolians from white men; presumably we might add, to differentiate Aryans from Semites. But, apart from a single reference to views held by Mr. A. R. Wallace, Bagehot does not enter upon this question at all. Granted race evolution, he asks how political evolution proceeds. Do we encounter in it the workings of inheritance and natural selection? If so, what forms do they take?

But even within the political region two problems are entangled together—if, indeed, I ought not rather to say that there are two different ways of conceiving the one political problem. This doubleness of aspect or of parts is embarrassing; yet it is a difficulty we often encounter as we follow evolutionary discussions, especially those which bear upon man. Does evolution mean progress, or does it simply mean differentiation? By wedding " Physics," *i.e.* biology, and " Politics," are we seeking to explain the cause of political changes or rather of political improvement? Parts of Bagehot's book deal with the latter point, especially his closing chapters. On the other hand, are we simply trying to explain the origin, from one common stock, of the immensely divergent assemblage of national constitutions which history records or living experience manifests? This question is also in his view. Perhaps we ought to say that he wishes to study both phases of his theme, but that he is chiefly interested in the laws of true progress.

Before history, he tells us, there was a prehistoric age, before morals, a non-moral age. If man was created, he must have had everything to learn. If man was evolved from purely animal forms—this Bagehot seems to regard as probable, but as non-essential to his argument — there must have been an interregnum between the time when instinct guided action and the time when reason became effective. Instinct on the whole secures safety, but reason weakens instinct, and custom, which is the equivalent of instinct at a higher grade, which is the earliest and most important safeguard of rational beings, must have been very slowly and very gradually formulated. Primitive savages were like modern savages in almost all their defects ; they were ignorant, capricious, passionate ; but their minds cannot have been " tattooed over with customs " like the minds of their remote posterity, the savages of to-day. While civilised man is social, primitive man, according to Bagehot, was a being no longer guided by animal instinct, but imperfectly human, and very hard to break to the sway of society. Most men were wild ; many races were purely wild ; and the vital problem during the emergence of society was to secure the formation of " a cake of custom " which might keep savage nature in check. Good custom or bad might serve ; the quality of the custom was a secondary though doubtless very important point ; its existence was the main thing. " Any sort of government was better than none at all." But in this, as in so many matters, the first step was much the hardest. Once he had laid aside his primitive rudeness, the imitativeness of man made everything easy. Imitation continued old customs, imitation diffused attractive novelties. It was thus both a conservative and a progressive force, but it was

oftenest at work in the service of inherited usage.
Here then were the factors of social order—custom and
imitation. Once the race became political it developed
an overwhelming power of conservatism. Custom had
made men what they were ; they dimly felt this and
worshipped every custom with equal enthusiasm, the
worst no less than the best. But indeed isolation was
useful in early days. Jealousy of novel or foreign
ways was a wise passion while the social type was too
weak to bear contact with other types.

In the way of comment or criticism one need only
here remark, that almost everything in this eulogy of
custom turns upon Bagehot's theory of the unsocial
wildness of the first men, or, as he tends to translate
that conception, on the theory that, when man was
evolved, instinct went off duty before reason and custom
came on duty. Probably that proposition is disputable.
And the whole attempt to affirm how reason *must* have
proceeded in entering a world that knew it not is
perhaps an attempt to transcend the limits of possible
knowledge, more truly so than many things which have
been thus described.

Custom being established, the next question to be
faced is, how the cake of custom may be broken and
progress inaugurated. Custom, and the rough natural
selection of early ages, ensure stability ; they are the
factors in social statics ; but what are the factors in
social dynamics ? For a long time the greatest selecting
agency is war. Military nations prevail over those
which are less effective upon the field of battle, and to a
large extent imitation gradually diffuses the principles
of the higher and conquering civilisation among the
vanquished. For in a sense the conquering civilisation
is higher. Reflection shows us that, up to a certain

point, the best man wins in the fierce competition of
war.   The military virtues are correlated to other virtues,
or they are serviceable for other ends besides conquest.
Beyond a certain point, however, progress is not secured.
War tests and develops the military virtues, but it does
nothing to hinder the heavy weight of custom from
crushing out the finer possibilities of human nature.
On the contrary, as we know from Mr. Spencer,
militarism is the natural ally of autocracy and of
reaction ; it calls for a blind obedience.   Therefore, to
end this paragraph as we began it, we are called on by
Bagehot to notice how very many civilisations have
become stagnant ; how very few have been the instances
of progress ; how many beginnings that promised well
have suffered a speedy arrest.   In the same spirit
another distinguished writer, Sir Henry Maine, has
taught us that the barbarian inroads may have been
needed to save Europe from the fate of China.   These
positions are memorable in view of what we shall hear
from Mr. Benjamin Kidd (speaking on the authority of
Professor Weismann) that for every organism the choice
lies between struggle, victory, and progress on the one
hand, and continuous retrogression on the other hand.
China has at least worn the appearance of stagnation
for many ages.   China seems to have evaded Mr. Kidd's
dilemma.

But, if war has a limited power of selection, and
effects a certain amount of progress, the decisive step
has been due twice over to the influence of free discussion in the sphere of government.   The habit of
political debate in the Greek democracies, the same
habit afterwards as a tradition of the Teutonic peoples,
kindled and enflamed the mental activity of civilised
men, till discussion, like a forest fire, had spread to all

the material within reach. Democracy is not needful for
this effect. The so-called Greek democracies were really
petty aristocracies of slave-holders. You may have as
high a franchise as you like, yet, if free discussion prevails within the privileged circle, then the emancipating
force is at work. Mere oratory may not educate. The
graceful oratory of the Red Indians dealt with methods,
not with principles, and effected nothing towards progress in civilisation. But, when political discussion
deals with great topics, it has a marvellously stimulating
and educating effect on the mind. That has been the
chief factor in social dynamics. That has twice broken
the cake of custom. And now the intellect is fully
awake, and progress itself has become a tradition of the
western world.

In subordination to these great factors Bagehot notes
others. For example, he dwells on the importance of
the blending of races. Such mixture, it is thought,
frequently improves the breed, and so leads to evolutionary progress. But even if it results in no improvement—or even if it tends to deterioration—it may
yield a new type, and so conduce to variety of result;
if not to progress, yet to differentiation.

We take leave then of this most interesting little
book with three remarks. First; it does not yet show
us Darwinism in relation to ethics or even in relation to
sociology in the stricter sense, but rather in relation to
politics. Now in politics there can be no question that
we have before us a spectacle of competition—pre-
eminently, but by no means solely, in the fierce rivalries
of actual war. And so the application of Darwinian
ideas in this region is unquestionably lawful, if a trifle
obvious. Secondly; in spite of his references to the
nervous system, Bagehot assumes inheritance mainly by

the psychical and political forces of imitation and custom.   Thirdly ; he does not to any great extent connect
the other side of politics—progress, social dynamics—
with natural selection in the strict sense.   Progress as
well as stability rests upon imitation and upon the
possibility of loans in culture.   To a certain extent progress rests upon war—but not upon wars of extermination ; not, therefore, on elimination of the unfit and
survival of none but the fittest.   Mainly progress is due
to the habit of political discussion, and to happy circumstances giving that habit great effect.   In other words,
Bagehot's social dynamics centre round a purely political
idea.   Not the biological analogy but special historical
knowledge has been his guide.   Darwin has set him
thinking, but Darwinism has not mastered or over-
mastered the course of his thought.   This is not said by
way of blame or disparagement, but in order that we
may reach a precise view of the nature of Bagehot's
contribution, and may understand how it differs from
other contributions with which we have still to deal.

NOTE B. *On Professor Ritchie's " Darwinism
and Politics"*

[Professor Ritchie's bright little book does not propose to apply Darwinism to the details of social life or
history.   It deals with the question whether the application holds good in principle—whether or not Darwinism
really applies to politics.   Unfortunately it is not easy
to harmonise the teaching of the different essays.   The
bearing of the first essay is as follows :—Whatever pre-
sumptions are established by a Darwinian view of the
origin of man, there is no ground for believing that social
progress necessarily implies struggle ; reason has come in

to change all things.   But the drift of Essays II. and III.
is in quite a different direction :—The analysis of evolution by Darwinism is absolutely trustworthy, and may
assuredly be extended to human society, " mutatis
mutandis ! "   This implies that reason has made only
minute changes.   Yet the first essay teaches that reason
has equalised the efficiency of the two sexes, and again,
that it has suspended the necessity for struggle.   How
much Darwinism is left if you eliminate struggle for
existence ?

There is one hint of some interest in Essay I.—that
reason, as embodied in governments, may achieve a
better *economy of material* than is done by [" natural
selection " or] *laissez faire*.   But whatever the value of
this hint, it is not Darwinian.   And the title promises
Darwinism ; and that is what we are studying at this
moment.

Some further remarks on Prof. Ritchie's positions will
be found in Chapters XVII. and XX.]

CHAPTER XIII

DARWINISM IN ETHICS—PROFESSOR ALEXANDER

Fusion of idealism and naturalism—Moral judgments are facts, but the assertion of free-will is absurd—Criticism ; capricious ; ignores the content of moral judgments and the germ of a system in them—Punishment grouped with dynamics ?—*Statics* are truly, though imperfectly, moral—Goodness is a twofold " equilibrium "—This doctrine is enforced against other definitions —In the *Dynamics* equilibrium is revealed as endlessly changing, and is called " compromise " — Ideals compete like organisms for survival —- Criticism ; not (*a*) true Darwinian struggle, nor (*b*) true extinction—The new ideals are not wholly new—Ideals are complementary—So far as he Darwinises he is false to morality.

PROF. ALEXANDER'S *Moral Order and Progress* is a very full, interesting, and original discussion. Its character, as the sub-title indicates, is " an analysis of Ethical Conceptions." The general position of the author is that of one struck with the convergence of idealistic and naturalistic ethics in the light of evolutionism; but, while coming himself from the camp of the idealists, Mr. Alexander is strongly inclined to seek a place in the left wing of the partially amalgamated forces. All that is true or solid in idealist ethics is provided for, he thinks, in the biological scheme. As for intuitionalism, it may go packing; there is no portion for it in the promised land of truth ; it is mere mischievous illusion. We have been told by some of Lord Beaconsfield's admirers that there was a great unity throughout his career, in spite of all apparent change—he always disliked the middle classes. Against them he appealed variously to the nobles and the poor, to Tory and Radical instincts. So it is to be with the typical *bourgeois* philosophy of intuitionalism. Idealists and empiricists are to agree sweetly in destroying it. Its excellent intentions shall not excuse it one cruel blow, in view of its hopeless and irritating limitations.

Having affirmed so strongly the competency of naturalism, Mr. Alexander has to face a question which, in our judgment, presses hard upon all naturalistic ethics. What room is there for ethics at all upon the premises of naturalism? What do we mean by speaking of right and wrong, of moral good and moral evil, in a world of blind laws and mere facts and necessary processes? Mr. Alexander, like Mr. Stephen, faces the question and gives the same provisional answer. Primarily, we are dealing with acknowledged facts, viz. with those moral judgments which, as a matter of fact, are current. In the first instance, therefore, Mr. Alexander takes over moral opinion as he finds it, and, like Mr. Stephen, tells us he is concerned to analyse it rather than to verify it—to systematise it, as we might perhaps interpret, rather than to apply any more radical test. Self-consistency is indeed a legitimate test, though but a negative test of truth ; and if he had confined himself to requiring that morality should be self-consistent, coherent, systematic, Mr. Alexander could have done no possible injustice to the moral consciousness. As we read on, however, we feel that his provisional attitude is very soon departed from. The utterances of the moral conscious-ness are cut short—its dicta are edited or expurgated— with a view to securing harmony, not with each other, but with a deterministic view of the universe borrowed from physics. True, the frontier of morality is extended

a long way in certain directions. With admirable faithfulness Mr. Alexander reports that conscience passes its
judgments on willed conduct—only on willed conduct ;
yet scarcely is this admitted when free-will is mockingly
expelled from the court unheard—free-will, the one
further truth which gives meaning and justification to
our human habit of passing judgment only upon will.
Why is free-will exiled ? What procured this order
from the judge ? Morality did not require it ; conscience
asked nothing of the kind ; victorious prejudice, and
the tyranny of physical science, carried the day. That
is not the way to provide our subject with a scientific
frontier ! It results in a haphazard frontier—pushed
far on, at one point, to suit the requirements of our
own position, but then cut short to suit the requirements
of other people across the border. Mr. Alexander is
loyal to the psychological fact that we judge only
willed conduct ; he takes care to report it accurately ;
but what does he make of it ? Stated in isolation, is it
not meaningless ?

We see now in how restricted a sense moral facts
are admitted by Mr. Alexander. The moral consciousness is allowed to bear testimony ; " AB is an ethical
conception " ; " CD is an ethical conception "—but that
is all. The authority of conscience is good to that extent—and not an inch beyond. If we ask the further
question, what is the meaning of this ethical conception
AB ? Conscience falters and grows embarrassed, or
remits the matter for analysis to the laboratory of
ethical science. From this point onwards conscience is
dumb, and Mr. Alexander acts as its proxy, or works
up, as he judges good, the material with which it has
furnished him.

This criticism must not be misunderstood. We

should not think for a moment of denying the rights
and privileges of reflection, or of questioning its value.
When moral opinion has done its utmost in the shape
of healthy instinct, very much remains to be learned
from the brooding meditative critic, who insists that
we shall " see life steadily and see it whole," and who
therefore brings our scattered thoughts into focus and
tunes them together as a harmonious system. When
that is faithfully done the moral philosopher is not the
tyrant, but the *minister atque interpres* of conscience,
carrying on its own work and giving it a higher perfection. He may indeed do more than this. He may
provisionally call in question the teachings of conscience ;
he may subject them to tests ; provided he recognises
that conscience has its own contributions to make to
any final synthesis. But all this describes something
very different from Professor Alexander's treatment of
the subject. We do not blame him for revising or
modifying the dicta of moral instinct, but for the kind
of revision he practices,—one which ignores that the
process of interpretation is begun by conscience itself ;
one which lays down the law upon questions of morals
in obedience to non-moral principles ; one which treats
the law thus laid down as decisive against the moral
claims of free-will. Conscience is invoked to supply
our author with facts for manipulation ; it is allowed
to do nothing more.

We cannot attempt to follow out Mr. Alexander's
interesting discussion in detail. We can only name
a few points which seem specially noteworthy, either
for their own sake, or in connection with the history
of the appeal to biology for human guidance.

The subject is explicitly divided into two main
parts—a statical and a dynamical ; moral order, and

moral progress; in obvious dependence upon Comte.
One must be allowed to express a doubt whether
names and things exactly correspond to each other
here.   As a point of detail, it is astonishing that
punishment should be discussed under moral progress.
If there is any obstinately statical element in the life
of society, surely it is penal law, which maintains
what has been reached, but is grimly indifferent to
further progress.   When saints or martyrs challenge
a law that has been outgrown, or that is downright
bad, there may of course be progress through the
punishment they bear—thanks to them, not to the law.
In itself the law does not even then make for progress.
Its preoccupation, then as always, is stability.   And the
ordinary victim of penal law is much more likely to be
affected by atavism than by " the prophetic soul of the
great world brooding on things to come."   What is he
doing in this galley ?

When one passes from details to principles, Mr.
Alexander's grouping of his materials looks more and
more disquieting.   He is really not contrasting moral
order with moral progress; he is giving us, first, an
analysis of morality in the abstract, apart from questions of progress, but secondly a theory of progress,
or rather of change, which sets morality at defiance.
In the first half—thanks to his appeal, however
strangely limited, to the moral consciousness—he is on
moral ground; the foundation is moral, whatever may
be the character of the superstructure.   In the second
half he has moved off moral ground altogether.   The
first is a theory of morality from the inside, if not
exactly from the heart of the subject; the second
is a theory of the changes in human opinion, a
view taken from the outside of the moral process,

and characterised by the airy indifference of the foreigner.

In Part I. the analysis of the moral end leads to the result that goodness is an equilibrium, and one of a twofold order. For *first*, goodness is an equilibrium among the promptings or desires or actions of the individual; and *secondly*, it is social, placing each man harmoniously with his fellows in an order of society. And this positive analysis is supported negatively by a destructive analysis of other views of the ethical end. To this extent therefore Mr. Alexander offers more *proof* in support of evolutionism in morals than Mr. Leslie Stephen gave us. Intuitionalism of course receives no attention. Intuitionalism holds that the good, like other primary elements of consciousness, cannot be decomposed, and neither can nor need be defined. It is hardly strange that one who is seeking a definition of the moral end should pass over such views in impatient silence. But, if intuitionalism is not discussed, a kindred position is faced when the definition of the end as *perfection* is brought under notice. This, says Mr. Alexander, gives no help. It carries us no further. Perfectly *what* should I be? Perfectly good, of course. But I am asking you what goodness is! You have told me nothing; you have taken for granted the conception of goodness. Next, Hedonism is discussed. Mr. Alexander dismisses as an over-refinement the idealist criticism, urged by T. H. Green or Mr. F. H. Bradley, according to which a *sum* of perishing pleasures is an impossibility. But he himself argues that pleasure cannot be the moral end, on the ground that there are ultimate irreducible qualitative differences between one kind of pleasure and another. Surely this does not seem altogether conclusive, especi-

ally since Mr. Alexander goes on to maintain that
his own formula incorporates hedonism by insisting
that *some* pleasures ought to be aimed at, viz. the
pleasures of goodness.   But there is no doubt that he
is right, from the point of view of the moral consciousness, in holding that if pleasure enters into the end of
[right] action, it cannot be pleasure as such but
desirable pleasure, *i.e.* morally desirable pleasure.
Lastly, Vitality is examined; and Mr. Stephen is
instructed that all that is true in this formula is
covered more exactly by the abstract formula, equilibrium.

So far as we have yet inspected this doctrine, it is
evidently akin to the older evolutionism of Spencer
or Leslie Stephen.   One organism, or one set of forces,
falls to be considered; goodness is a harmony in the
organism or among the forces ; badness is disharmony.
At first sight one thinks that Mr. Alexander has
materially improved upon Mr. Stephen's position.
With Mr. Stephen, the individual man and the social
whole fall violently asunder.   But Mr. Alexander knows
of a twofold moral equilibrium, applying alike to man
and to society.   Also one observes the traces of
Mr. Alexander's idealist schooling.   For him, morality
is still self-realisation or self-fulfilment.   Unlike intuitionalists, he regards goodness not as something
added from outside to the natural motives of men,
but as the correct working up of the raw material of
character.   It is true, Mr. Stephen, with his purely
empiricist tendencies, has caught the same truth.   But
the truth deserves full acknowledgment wherever found.
Assuming, as we are led to do, that the disorders in
character are many, the order, only one, there seems no
reason why we should quarrel with Mr. Alexander for

speaking of equilibrium as the moral end, if he likes to do so. Following his own lead we might hint that a different formula did fuller justice to the real contents of the moral end; but we should not condemn his formula as false.

A very different light, however, is thrown back upon this definition from the second part of Mr. Alexander's treatise. In it we learn that there are many competing and successive types of morality— endlessly many. Goodness is not one, in contrast to the multitudinousness of evil and disorder. Goodness itself is no less protean. We must not hold that morality is *the* equilibrium of conduct; each type of morality is *an* equilibrium. Without forestalling our discussion of the theory of moral progress, we notice now the bearing of this assertion not simply on the theory of moral order but on the very definition of morality. It had been proposed that we should define morality as equilibrium. That definition is now robbed of its meaning. Is there any conduct at all which may not be said to seek *an* " equilibrium "—if only that of the simple equation, " Let me be on the top and every one else below " ? Matters are not improved but rather made worse when the word " compromise " slips out as a synonym for " equilibrium." Is not almost everything a compromise—from some point of view ? The extortioner, the slayer of human lives, the cheat, " when he thinks of his opportunities," may, like Clive, be "astonished at his own moderation." You and I both claim something; half to me and half to you is a compromise; but ninety-nine per cent to me and one per cent to you is also a compromise. I may even persuade myself that a hundred per cent to me is a compromise, because I suffered you to get away with unrifled

pockets.   What possible light then is obtained by naming
good conduct "*a* compromise"?   A further objection
remains.   "Compromise" is the worst possible word for
describing moral behaviour.   Morality, as Mr. Alexander
bears witness, imposes a law, and that law requires uncon-
ditional obedience.   If we follow it out, our own nature
will blossom into its true richness and fulness; but
for this the knife is as necessary as the watering-can;
the path to moral self-development lies through self-
sacrifice.   Where is there room for talking of compromise in such a process?   The law indeed gives his due to
each man, and also to each impulse.   The "stern lawgiver" wears "the Godhead's most benignant grace";
but no wrangling of private interests, no arbitrary
delimitation of incompatible claims, will produce
morality.   In a word, morality involves order, equilibrium, peaceful settlement of competing claims; but
equilibrium — and still more plainly, compromise —
neither includes nor leads to morality.   Seek the
higher and the lower will be added.   Seek the lower
—you lose all.   We conclude therefore that Mr.
Alexander's theory is neither true nor false but merely
vague.

The second half of the treatise deals with moral
progress.   The most interesting and novel part of this
discussion is found in a doctrine laid down when
treating of the origin of moral distinctions; but, as
there seems to be no reason why the doctrine should
only be applied to the beginnings of moral progress,
we shall treat it as covering the whole field.   It sets
before us a vision of competing moral ideals, and of
the survival of the fittest.   The process is illimitable;
there is no absolutely best; every good, while it
is valid, or to those for whom it is valid, is also

the best; and as continuous evolution and adjustment go on, the moral ideal must vary or be renewed in correspondence with the facts of human progress. This assertion is treated as showing us the prolongation of the Darwinian struggle into new and higher regions. If men do not habitually struggle against each other, to the point of extinction for the vanquished and solitary survival for the victor, ideals do so; and the "creed outworn" succumbs, while the ideal which is up to date survives and predominates—for a season. So it always has been, so it always will be.

Such constructions of ideas seem very much akin to primitive mythology. Here too we have a metaphor, and here too the speaker does not know or does not remember that it is a metaphor, but treats it as a revelation of absolute scientific truth. The author uses most of the implications and inferences connected with Darwin's analysis, and uses them with dogmatic confidence. He never fully inquires what limits attend their use. Of course, it is possible to represent progress in thought as due to a competition between various types or ideals. Let us grant this in the fullest way. Such language is lawful; it may be suggestive and valuable. But metaphors are treacherous things; they leave out at least half the truth.

Natural selection takes place, or is alleged to take place, through the struggle for existence, because there is not room for all to live and be nourished side by side. Every living organism *cannot* live out its full time and transmit its peculiarities to offspring. But what forbids moral ideals to exist side by side? Truth to tell, they have done so in the past, and do so yet—in different lands, or even in one land—in different minds, or even in one mind. The struggle of ideals is much less keen than the biological struggle for existence, at least at starting, and in its lower stages ; afterwards its working may become swift and telling. Ideals compete against each other in human minds. They commend themselves not by any physical superiority, but by their attractiveness or by their truth.

Secondly, there is a difference mentioned by Mr. Alexander himself. Defeat here, in the struggle of ideals, does not imply the extinction of the persons holding inferior moral conceptions. The ideals perish ; the persons who held them are usually converted to a higher way of thinking. Surely here we have an open admission that the struggle between ideals is not a struggle of the Darwinian order. Progress according to Darwin *is dependent on the weeding out of the unfit.* Progress according to Mr. Alexander is usually secured by a conversion from error to truth. It is a secondary result that errors disappear. And those who were formerly in the grasp of error do not die, but believe the truth and live.

Yes, it may be said, the errors die. Is not that enough to justify the analogy ? Let us look then a little more closely at the alleged mechanism of moral progress. Variation constitutes, says Mr. Alexander, a *new species* or *new ideal*, before which, after a season of struggle, old species or old ideals perish. Does not this statement ignore the fundamental continuity of life throughout all evolution ? The " new species " *is* an old species modified. The new ideal is not wholly new ; it is the fuller evolution or unfolding of the old, what Hegel called its truth.

For of ideals above all things we may declare that they do not struggle blindly against each other, or exclude each other. They are not physically distinct things, mutually incompatible, mutually repulsive. Was there ever an ideal with a lower programme than that of the supreme Teacher, "Not to destroy, but to fulfil"? The point may be illustrated by a quotation from John M'Leod Campbell: "An early member of the Society of Friends, writing to a brother who was a Roman Catholic, says, 'Your religion and my religion must be the same, in so far as we have religion, for there is but one religion.' This true and deep word," adds Campbell, "we are gradually learning to understand." May we not even more confidently say the same thing of moral ideals? There is but one ideal. The various forms in which, historically, the ideal presents itself are not distinct and rival species, but elements in the final synthesis—yearning aspirations after it—sketches, rough and rude at the best, yet instinct with life, and all representing one great pattern seen in the mount. Would an ideal kill another ideal if it could? I do not ask, would an idealist kill an idealist? That indeed is "another story"; but does the ideal itself aim at extermination and destruction? Mr. Alexander tells us that the rivals often blend in a "compromise." Surely, once again, the victory of truth is no compromise between opposite extremes, but something higher than either, in which all that is best in both the rivals lives on and flourishes. And the *tertium quid* at least *may* be due to a victory of truth.

We conclude then that the application of Darwinism to competing moral ideals breaks down all along the line. For, first, what is described to us is not a process of natural selection by means of a struggle for existence; and, secondly, so far as Mr. Alexander does assimilate moral ideals to competing organisms, he falsifies the

facts.   He has not really shown us an extension of
Darwinian struggle into a higher region, but something
radically different—something described by him more
or less suggestively, but also more or less inaccurately,
in Darwinian language.   Progress by struggle — *this*
morality thrusting down *that* morality and reigning in
its stead—is not exhibited in the facts of history to any
one who can look ever so little below the surface.   Moral
progress is much better described from Mr. Stephen's
point of view as one great orderly evolution of human
thought and life.   Mr. Alexander sometimes uses similar
language ; but if such language were meant in full
earnest it would be necessary to cease speaking of the
limitlessness or indefiniteness of moral change.   *We*
may be baffled and bewildered by the course of moral
evolution.   Many a time good but timid men have
regarded change and even advance in moral conduct or
ideas as pure wanton iconoclasm.   But it was not so ;
it was inwardly continuous with what went before.
And, although philosophy itself must fail if it seeks to
forecast the morality of a distant future, yet the future
form will grow out of the present, and, when it comes,
men will see in it once more how wisely and how surely
God fulfils Himself.   To abandon that hope is to
abandon morality and all that makes us human.

CHAPTER XIV

Reaction as to ethics—Due to the vision of *struggle* and *pain*—Not sympathy, but justice is essential—It must suspend *outright* the cosmic process—Older evolutionism (Greece, India) gave no guidance—Criticism : nature and spirit are opposed—Yet connected, and reason fulfils the cosmic process by transforming it.

IT will readily be divined that it is in a special sense we connect the name of Huxley with reaction from Darwinism. From the time when he was converted to the new views, Huxley was perhaps their most brilliant and successful advocate, both in scientific circles and as a populariser, speaking to the world of readers. Yet, in regard to ethics, he was continually restive. The Romanes lecture for 1893 is only the most deliberate among many striking utterances of his, tending in that direction. His thesis runs to the following effect, that evolutionary science has done nothing for ethics ; that, on the contrary, men only become ethical as they set themselves against the principles embodied in the evolutionary process of the animal world. Far from regarding evolution as the master-key to ethics, Huxley insists that the two terms are irreconcilable.

Plainly, Huxley has considered only one possible form of union between evolution and ethics. For him evolution means Darwinism ; the struggle for exist-ence which is believed to have dominated the plant

and animal kingdoms.   And for him the union of
evolution with ethics means not analogy but identity;
it means that man, the individual organism, is held to
become moral by succeeding in the struggle for existence—a sufficiently startling paradox.   Huxley makes
no explicit reference to Spencer's formula, tracing a single
harmonious process, right back to the primeval nebula
and right on to moralised man.   He is willing to gener-
alise evolution as much as you please, but it seems to him
that there is a seriously novel element introduced at one
point in the process, cutting it as it were in two. " When
the cosmopoietic energy works through sentient beings
there arises among its other manifestations that which
we call pain or suffering."   And suffering is most intense
in man, especially as he rises in the scale of civilisation,
"under those conditions which are essential to the full
development of his noblest powers." [1]   Animal struggle
runs on into human struggle, but such struggle is
immoral.   We must not wantonly add to the pain
suffered by our fellows; we must " let the ape and tiger
die."   The Spencerian formula—so we may read between
the lines—makes no room for those elements which, to
Huxley's mind, are of real moral significance.   As for
Comte's attempt to view social life as the evolution of
one orderly and peaceful organism, or as to Mr. Leslie
Stephen's gloss upon that attempt, or as to Professor
Alexander's bloodless and well-nigh painless Darwinism
in the shape of competing ethical types, Huxley says
nothing.   He cannot separate evolution from the cruel
Darwinian struggle in its plain and literal sense.   He
puts ethics and evolution as far asunder as the poles.
We might almost style him a valuable if unexpected
recruit to the cause of Miss Frances Power Cobbe.

[1] p. 10.

Darwin of course he knows by heart; and Darwin's easy-going ethics felt none of his difficulties. How does he answer Darwin's proposal to deduce morality from sociability *plus* intelligence ? Primarily, it would seem, by emphasising *justice* as the moral ideal rather than sympathy. Sociability might conceivably explain the rise of sympathy, but not of a sense of justice. " Wolves," he says, " could not hunt in packs except for the real though unexpressed understanding that they should not attack one another during the chase. The most rudimentary polity is a pack of men living under the like tacit or expressed convention ; and having made the very important advance upon wolf society, that they agree to use the force of the whole body against individuals who violate it, and in favour of individuals who observe it." Out of this convention arises a sense of justice, within the human pack ; and justice is gradually deepened into *righteousness*. Now certainly such a conception of the moral ideal is not so easily fitted on to an evolutionary process as a more purely altruistic conception of goodness. Darwin thought sympathy or comradeship the chief point in ethics. Huxley swears by justice. He is tempted to call nature unjust ; he is sure that it is non-just.

Once again, in a note, he returns to this point. Having by that time formulated the evil of cosmical nature not simply as pain, but as competition or struggle, he adverts to the fact that packs of wolves, hives of bees, and all social or gregarious creatures have suspended the struggle within their own community. " To this extent," he admits, " the cosmic process begins to be checked by a rudimentary ethical process, which is, strictly speaking, part of the former, just as the ' governor ' in a steam-engine is part of the mechanism

of the engine." [1]   This represents the sum total of the
concessions which he would make to those like Messrs.
Geddes and Thomson or the late Henry Drummond,
who allege that Nature is not wholly red in tooth and
claw, but that a principle of love is gradually disclosed
and made predominant as we ascend the evolutionary
scale.   He grants that the wicked process of struggle
is partially, slightly, very slightly checked, and checked
by justice ; but, in the main, cosmical nature is full of
struggle, and, from our human point of view, full of
wickedness.

The rest of the lecture does not add very much to
these essential ideas.   It verifies them by tracing former
evolutionary thought in India and Greece.   Indian
wisdom regarded all things as embraced in an evolutionary process extending through æon after æon, and
life upon life ; but it held this process to be downright
bad and unhappy.   Buddhism, its most characteristic
expression, rested on a pessimistic view of the world ;
such pessimism may have been one-sided, but its existence proves how little a belief in cosmic evolution did,
in those days, to guide men as to their personal conduct.
The cosmic process said " Live ! "   The enlightened one
said " Extinguish yourselves ! "   In Greece, the ethic
of the Stoics was alleged to be connected with their
Pantheistic evolutionism ; but Huxley contends that it
was really perfectly independent of its speculative background ; and that is very likely true.   Coming down to
modern times, he complains that discovery of " the
evolution of ethics " has led men, in much confusion of
thought, to preach an " ethics of evolution " ; whereas
no such thing exists.   Good of course has been evolved
—but so has evil ; beauty has arisen in evolution—and

[1] p. 197.

ugliness too ; what survives after struggle is "fittest to survive," but not necessarily best or noblest. Briefly, cosmical and ethical tendencies are opposite. We human beings have to develop our own ideas of justice ; the bad blind world can neither guide nor help us. In the past, struggle was of service when it gave man dominion over the creatures (as theologians express it)—a curious hint. But now the remainders of struggle poison man's higher life.

Perhaps this is seasonable discourse. After all, nature and spirit are different things, and, if philosophy drops below pantheism into downright materialism and atheism, then too probably it will undermine morality. Nevertheless we must not exaggerate the difficulties of the case, or leap prematurely to the sorry conclusion, that nature is in opposition to morality. We are not obliged to rush into either extreme. Because we hesitate to recognise evolution as the key to ethics, we are not bound to regard evolution as anti-ethical. Huxley seems very one-sided when he draws a sharp contrast between the best and those fittest to survive. Bagehot and Mr. Leslie Stephen teach a different lesson. Among human societies it is probably fair to assume that in the majority of cases the most moral are the strongest. So far as that is true of states or of individuals, the "blind" cosmic process does not oppose morality, but acts in its service. The difficulty is at least attenuated.

A fuller answer to Huxley's perplexities regarding the moral bearings of evolution is to be found in a better view of reason. Morality is a new thing in the creation with the advent of rational man, yet not wholly new. It is the *transformation* and perfecting of animal ethics —not the simple inversion of the cosmic process. But it is a highly significant transformation. Pain also is

transformed by the advent of reason. Even in the
animal world, presumably, pain is outweighed by
pleasure, Huxley himself being witness. In man, however, pain assumes a new meaning. It becomes an
element in moral development. How then can the
presence of pain brand the cosmos as evil? The
kindred charge, that struggle is altogether evil from
the moral point of view, will come before us again in
the next and subsequent chapters; we trust there are
reasons for repelling that charge also. Lastly, we
observe that a more intelligent conception of reason
corrects Huxley's position as to the supremacy of man.
Mastery of the animals is natural to mankind. It is no
mere accident, due to man's share in the cruel cosmic
struggle. It is man's right. It forms part of his
equipment for that which lies before him,—the moral
struggle to which the cosmic struggle gives place, the
moral advance and moral achievement which are to
crown the long and strange story of this earth.

CHAPTER XV

REACTION FROM DARWINISM—DRUMMOND'S
"ASCENT OF MAN"

His precursors—His sympathy for Spencer—His Comtist terminology—Seeks a
biological basis for altruism—Corrects Darwin—Not like Miss Cobbe—
Largely like Huxley—But seeks a fairer statement of the facts—Brings in
a second biological function (out of three !), viz. reproduction—Wallace on
the selection of reason—Leads up to doctrine of "*Arrest of the Body*"—
Cf. Clelland on the human skull—Emphasis on *maternity* and weakness of
human infant—Criticism ; "egoism" and its struggle purely evil ?—Or
male sex with its justice ?—Is domesticity = sociality ?—Has Drummond
shown a *factor* in progress ?—A better philosophy claims *all* nature for
God.

I HAVE chosen the *Ascent of Man* to represent the more
conscious and definite reaction from unmodified or un-
balanced theories of natural selection, not because its
author was the first or the only writer to champion such
a reaction, but because he has given us its fullest statement, and because everything of Drummond's com-
manded at once a very wide popularity. For another
reason he interests us, because he speaks as a Christian
believer and thinker,—almost as a Christian apologist.
He himself confesses obligations to many predecessors ;
first, perhaps, to John Fiske, as we shall note in due
course ; most largely and definitely to *The Evolution of
Sex* by Professor Geddes and Professor J. A. Thomson.
These last writers, like Drummond, are consciously
dissenting from Darwin,—consciously putting forward
amendments to his statement of things, and not only

to his statement of the basis of morals, but to his scientific formulation of the process of evolution itself. Morality is to be found somewhere in the region of sex. Struggle for life is a fact, but not the whole fact; it is balanced by struggle for the life of others. Yet those who so speak are themselves evolutionists,—themselves Darwinians. They accept struggle for existence as a great fact and potent cause of progress. They deny it to be the only fact; and occasionally they are found denying that it is the only cause of progress; but that topic is very lightly touched upon. Hence perhaps, in part, one's perplexity, when one seeks to estimate the value of this correction of Darwin's theories.

With the wider Spencerian doctrine of evolution Drummond takes little to do. Yet he seems to assume its truth, or the truth of something of the same nature. His lyrical outbursts of praise at the thought of evolutionary science refer to something much more extensive than any view of the origin of species. Speaking of "evolution in general," he tells us that "Evolution is a Vision, . . . which is revolutionising the world of nature and of thought." When the workers of science had whispered the name "Evolution," "henceforth their work was one, science was one, the world was one, and mind, which had discovered the oneness, was one."[1] Again somewhat later we read, "Nature in vertical section offers no break or pause or flaw." To study it in horizontal section "is to study a hundred unrelated sciences—sciences of atoms, sciences of cells, sciences of souls, sciences of societies; to study it vertically is to deal with one science—evolution."[2] All this points to Spencer's philosophy, or a cosmic philosophy of a similar type. Yet such a system is nothing but ornamental

---

[1] *Ascent of Man*, p. 1.   [2] *Ibid.* p. 59.

scenery, hung up in the background of Mr. Drummond's *atelier*. His references to it during his discussion are of the slightest. Close to the end of his book[1] there is a whimsical attempt to trace the cosmic principle of love down into the inorganic world, and back to the nebulous cloud out of which natural law is said to have evolved all things. Chemical affinity is the supposed representative of the psychical principle of love, grouping the elements of nature in close union! However, the author does not seem perfectly easy in his own mind as to this suggestion, or thoroughly in earnest with it. On at least two other occasions he quotes Spencerian language in a tone of discipleship. "The first work of evolution always is, as we have seen, to create a mass of similar things—atoms, cells, men; and the second is to break up that mass into as many different kinds of things as possible. Aggregation masses the raw material, collects the clay for the potter; differentiation destroys the featureless monotonies as fast as they are formed, and gives them back in new and varied forms."[2] Again : "According to evolutional philosophy there are three great marks or necessities of all true development —Aggregation, or the massing of things; Differentiation, or the varying of things; and Integration, or the reuniting of things into higher wholes. All these processes are brought about by sex more perfectly than by any other factor known."[3] Except for these passing salutations, however, there is no appeal to the laws of physical or sub-organic evolution. We are bidden indeed follow nature; we are bidden throw ourselves into the current of evolution; but it is animated nature that is to be our guide; the nature which Darwin

---

[1] *Ascent of Man*, p. 433. [2] *Ibid.* p. 320.
[3] *Ibid.* pp. 336, 337.

studied will teach us rightly—if we a little readjust the formula in which Darwin summed up his results.

Going back a step farther, from Spencer to Comte, we cannot but be struck with the extraordinary closeness of discipleship manifested by Drummond. If Comte started the process of naturalistic study of duty under the flag of sociology, Drummond accepts the whole programme. The appeal to history disappears; with all his varied culture that was not in Drummond's line. But the appeal to biology stands; the conception of altruism as a synonym for virtue stands firm; the conception of sociology as an authoritative science, growing out of biology, is accepted in so many words. "Every earnest mind is prepared to welcome" sociology, "not only as the coming science, but as the crowning Science of all the Sciences, the Science indeed for which it will one day be seen every other science exists. What it waits for meantime is what every science has had to wait for, exhaustive observation of the facts and ways of Nature. Geology stood still for centuries waiting for those who would simply look at the facts. . . . Sociology has had its Werners; it awaits its Huttons. The method of sociology must be the method of all the natural sciences. It also must go and see the world making, not where the conditions are already abnormal beyond recall, or where man, by irregular action, has already obscured everything but the conditions of failure, but in lower Nature which makes no mistakes, and in the fairer reaches of a higher world, where the quality and the stability of the progress are guarantees that the eternal order of Nature has had her uncorrupted way."[1]

Most noteworthy perhaps, in comparison with Comte, is the attempt to justify the definition of virtue as

_____
[1] *Ascent of Man*, p. 57.

" altruism " by some biological considerations. We shall speak more in detail of this presently. If it should stand, would it not be another great stroke of luck for Comte? or, ought I to say, a further vindication of his prophetic insight? He did not foresee the evolutionary doctrine of the origin of species; he even deprecated such theorising. Yet the inquiry has gone forward, and the doctrine has been promulgated, and has set everybody using biological language. So too Comte did not think of justifying his favourite virtue of altruism by his favourite science of biology; yet that also has now been tried; and if the views for which Drummond is champion hold their ground, that also will have been accomplished. One can only repeat once more that it is extraordinary to find a Christian thinker such as Drummond casting in his lot so unreservedly with the programme of naturalistic science.

It is from Darwin, however, that the new discussion takes its departure. Its divergence from Darwinism is, in its own opinion, its most important feature. Let us look then for a moment at the peculiarities of Darwinism. All living species have been marked off from each other, and given a standing ground in nature, by the working of natural selection upon minute and apparently casual variations. The means of selection has been the ceaseless process of struggle for existence. At a certain point in this evolutionary process we have foreshadowings of morality when gregariousness appears, and when social sympathies begin to claim a place in animal life. Such limitation of the struggle for existence marks the dawn of morality. Henceforth sociality has only to develop its latent powers, and to call in the strong help of intelligence, and we have morality full blown. However, the struggle for existence is not terminated; it is only limited or modified.   Competition does not go on within the social group; "dog does not bite dog"; but the groups still compete with each other.   Morality and immorality are both of them natural products.   Evolution yields them both; they are both with us to this day in the strangest blending.   Darwin, being neither philosopher, nor moralist, but a student of facts and a seeker of natural laws, was content to publish his views of origin and process without inquiring very deeply into the probable consequences of such views in their bearing upon morality.

The first objection taken was by Miss Cobbe, speaking as an intuitionalist.   She complained that morality had no more sacredness, no more binding force, if it were true that conscience was a simple remainder of brute tendencies, useful to the species, but having no ideal sanction.   That objection we have ventured to overrule.   Provided only a sufficiently deep view of intelligence or reason be held—provided we see clearly that reason transforms, perfects, makes new, what it seems to inherit from brute nature—we need not be afraid for morality though it should universally be taught that morality came into being by slow and gradual fashioning of brute impulse.

A somewhat different objection is in the view of Huxley and Drummond,—not the origin of conscience, not the inheritance of moral instinct from brutes, but the swamping (as it were) of moral instinct in the great current of cosmic process, regarded as a struggle for existence.   If all nature struggles blindly and selfishly, what should man be but a "strugforlifeur" like the villain in Daudet's novel?   If reason, so we may interpret the difficulty in the light of Mr. Benjamin Kidd's work, the destined goal of our present study—if reason teaches man that the whole animated cosmos has been and is controlled by a struggle for existence, and by that struggle has been pushed onward and upward, what can man do but reverently bow down before blind selfishness, and practise it in his own life? Mr. Huxley, a man of science among the moralists, a Saul among the prophets, advancing boldly like Athanasius *contra mundum*, preached the absolute opposition of human morality to cosmic process, and called on his fellow-men to be moral in spite of the nature of things, the cruel, selfish, pain-dealing nature of things, from which we of the human race have arisen.

Mr. Drummond and others agree with Huxley and the "strugforlifeur" as to the effect of Darwin's views. But they argue that Darwin's views are one-sided. They ask us to define nature more exactly. And they fall back upon biology and its categories in making their new survey of the cosmic process.

Biology, they tell us, has two main functions, nutrition and reproduction. There is indeed a *third* biological function, corelation; but no account is taken of it, in order "to avoid confusing the immediate issue";[1] surely a rather airy fashion of dealing with the authority of science? It is indeed hard to see how and where the omitted function is ever to gain a hearing for itself in the new ethic, based upon the true biology. For the two functions already in evidence seem between them to cover the whole ground. "Nutrition" and "reproduction," the "hunger" and "love" of Schiller's witty stanza, claim the whole of life as theirs in joint tenure. The struggle for existence belongs to the first function; it is a struggle for nutrition; reproduction, with its "other-regard," manifests itself in struggle for the life of

[1] *Ascent of Man*, p. 17.

others.   The male sex stands for the first; the female sex
for the second.   Out of the one arises egoism; out of the
other altruism.   In their lowest germs these two physiological forces are held to have in themselves and to
make manifest the prophecy of their final moral result.
Even in reproduction by fission, when a low organism
overtaxed by the claims of nutrition upon its existing
surface splits in two and becomes two organisms,—even
there Drummond thought he could see the Divine law of
sacrifice worked into the very fabric of the animal world.
But without pressing such doubtful points we find him
urging that sociality and self-sacrifice grow more and more
manifest as evolution makes farther and farther advances,
a plain revelation (he thinks) that morality, the perfecting
of "altruism," is the goal of the entire cosmic process.

There are two points of special interest in Drummond's statement of evolution.   We may dwell shortly
upon both.   Even if the first does not directly elucidate
the alleged new conception of the evolutionary process,
it is important in connection with views that have still
to be considered.

The point in question is styled by Drummond "the
arrest of the body."   It seems to follow upon a contribution of Dr. A. R. Wallace's, which is very highly praised
by Mr. Fiske.   In answer to the question, How was
natural selection able to differentiate the rational species
of mankind from the brute tribes? or Why did not
reason die out as soon or as often as it emerged?   Dr.
Wallace replied that reason was preserved or was selected
*as soon as it became sufficient in amount to constitute
a greater advantage in the struggle than any physical
superiority.*   Upon this a previous question may arise.
How was reason, hitherto unfavoured by the selecting
agency, able to leap to that point of magnitude and

importance? That is a difficulty which besets the doctrine of natural selection all along the line, unless the admission is made that variation may proceed *per saltum.* However, in regard to the origin of reason, the difficulty is met *tant bien que mal* by treating reason alternately as identical with animal intelligence, and as something wholly new. When the origin of human reason is made the subject of discussion, it is spoken of as a new and advantageous variety; when the difficulty of its *quantity* or *amount* is referred to, it is treated as a slight improvement upon those lesser amounts of intelligence which are found among the highest of the lower animals. The muscular ape survived the feeble ape, and the clever ape survived the stupid one. The ape which was muscular but stupid, and the ape which was clever but feeble, ran perhaps a dead heat; but both of them were distanced a great way by the ape which was at once muscular and clever. At last, however, from one of the clever apes was born one cleverer still, one that deserved to be called rational, to be called human. And henceforth the future lay with him. He might be healthy or he might be feeble, but his endowment of reason made him more than a match for all the apes, more than a match for everything, unless another human child of the apes was evolved, who had the advantage of being more vigorous than the first, while equally rational. In that case the new-comer must be king! Of the two endowments, however —and this is Dr. Wallace's point—reason is the stronger. As soon as reason has become the thing best worth preserving by natural selection, rational beings survive. As soon as a rational race establishes itself, we may be sure that reason is the most important of all its helps in the struggle for existence.

To this contribution of Mr. Wallace's Drummond
adds the remark that the advent of reason involves the
arrest of the body.   Natural selection, it has been implied, is turning its attention to the mind.   Drummond
asks us to consider how this affects bodily evolution.   It
will terminate physical or animal progress.   Man has no
more need of an improved body ; he uses improved
rational methods.   In particular he supplements his
body by the use of tools.   But if man adds new
resources to the resources of his body, he also
counteracts many of its defects, *e.g.* he counteracts
defective eyesight by the use of spectacles.   There is a
danger here ; for it is implied that natural selection
does not kill off defective human types as it kills off
defective animal types.   We shall even be told by
Weismann that, natural selection ceasing to operate, we
ought to postulate not merely the arrest of the body,
but its retrogression.   Man might not retrograde as a
whole ; body *plus* reason he might become a more
effective creature in civilised times than he was in
savage or barbarous ages ; but what of his body ?
Confessedly, its advance has been arrested.   Is it not
inevitable that it should have receded, as civilisation
has been developed by reason ?   If we tried to verify
this suggestion by a reference to facts, we should probably meet with a good deal of evidence on both sides.
Except the few professional athletes, civilised men are
poor creatures physically in comparison with the higher
savages.   Whole faculties have gone amissing, and others
have left the merest aborted remnants.   Yet the civilised
man displays much physical toughness in the ordeal of
disease, while the " noble savage " breaks down.

Before leaving this point for the present, we ought
to refer to its bearing on the question of man's place in

nature. Is man the highest possible product of ter-
restrial evolution? That is plainly affirmed by Mr.
Fiske; and the same view is supported by Professor
Cleland of Glasgow,[1] on more specially anatomical
grounds, viz. that the human skull has been modified
absolutely as far as is possible in favour of brain. If
the "crowning race" wish to have much larger brains
than the Europeans of to-day, they must do without
noses, which would be very awkward for them, not
merely from æsthetic considerations.

The second point of special interest in Drummond's
statement is the "evolution of a mother." While sex
is the region in which morality is supposed to be con-
centrated, and while the female sex stand for goodness
and altruism in contrast to male egoism and badness,
Drummond makes it plain that morality first shows
itself not in love for the mate, but in love and care for
the offspring. That is true for the mother; in course
of time it becomes true for the father. Eventually
romantic love between the sexes comes as a long-delayed
climax. Rather sentimentally Drummond points out
that even plants are classed scientifically by a reference
to the reproductive process; that all the finest foods,
milk, fruit, grain, occur in nature for the sake of
reproduction, either animal or vegetable; that the
highest animals are named from the function of the
highest physical motherhood, *mammalia.* More noteworthy is the argument, originally Fiske's, that the still
higher development of human society, and with it of
human morality, is due to the feebleness of infancy.
The prolonged helplessness of human infancy kept the
family together, and gave depth and constancy to family
relationships. What again was the reason for that

---

[1] As cited by Drummond, *Ascent of Man*, p. 144.

helplessness of babyhood? The complexity of the processes gone through by an adult brain in rational life.
Animals, even the highest of the lower animals, have
comparatively few lessons to learn. Their nervous
system is always repeating the same combinations.
These grow stable by habit, and the young creature is
early emancipated from the care of its parents. Not so
is it with mankind. Here the elaborate education of
the nervous system *must* be a slow process. During
its long course pity, tenderness, love tremble into
consciousness; a mother is a mother indeed; man is
growing human.

Such in outline is the theory. What are we to say
of it?

Does Drummond mean us to understand, like Huxley
or like the Socialists, that struggle is purely bad in the
ethical region? Verbally, he denies this. It is " *struggle
for the life of others,*" not absence of struggle, which is
more and more to prevail till it dominates humanity.
Partly this struggle may be explained as carried on
against the forces of nature. Must it not also in part
be a struggle between group and group, home and
home? The struggle will no doubt be carried on
according to the laws of the game, those laws which we
know as justice. It will be lighted up and made digni-
fied by sympathy, by love for those within the group,
by consideration even for rivals without. That is a very
worthy programme. But does it not involve dropping
the old hard false opposition between egoism and
altruism, and dropping the somewhat apocryphal biological deduction of these two opposite tendencies? If
struggle is good, is there not an eternal use and fitness
in a limited amount of egoism? Or rather must not
that which is called egoism, and marked under that

name with obloquy, enter, however transformed, into the final moral constitution, and the highest human type?

Again we ask, can the male element be purely bad? And when we come on to the "evolution of a father," we find qualifications introduced. Rather to his own surprise, Drummond has to admit that the alleged feminine soul of goodness is not the only moral type. Authority has a place as well as tenderness; justice, or righteousness, Huxley's favourite virtue, is a specially masculine addition to the sympathetic virtues. Good again; but again tending to discredit Drummond's Comtist phraseology and his *quasi*-biological deduction of righteousness and of sin.

Another objection has been brought forward by Mr. B. Kidd. Drummond is said to confuse sociality and family affection, whereas they are distinct things. This seems of small importance. Probably the two things ought to be distinguished. Yet they co-operate; and, as Drummond has observed, the family is the strongest socialising influence.

We touch on a rather more serious point when we inquire whether "struggle for the life of others" is or is not a factor in physical progress. Once, but (I think) only once, Drummond deals with this question, and gives an affirmative answer, in so far as this, that the best mothers will rear the strongest and most successful offspring. Usually, however, morality or "altruism" is spoken of not as a cause or factor in evolution, but as a feature or result of the evolutionary process. The retort is almost inevitable from the side of pure or ultra-Darwinism, that natural selection by struggle is the whole fact, struggle for the life of others only a part of that fact, signifying struggle of group against group, yet assuredly signifying, still and always, struggle. If it be true that ultimately the whole race is "a moral organism," that "we are members one of another," that the highest and most advanced *need* the welfare of the most backward, that fact is a spiritual truth. We must not look to find it in nature ; we must not localise it in part of nature, and call this the moral part in contrast to the remainder, which is immoral or wicked. Nature is the presupposition of reason and morality, but reason and morality work up the whole of nature's raw material, not the half merely.

As against Huxley, Drummond seems to have been right. As against Darwin, he did not formulate any scientific difference. The same facts are in the view of both—the same facts differently stated and emphasised. To make a decisive advance, Drummond needed a more adequate philosophical schooling. He intended to vindicate all nature for God. Constantly he seems to be vindicating only a section, though perhaps a growing section. That position is of no possible interest to Christian theism.

CHAPTER XVI

REITERATION OF DARWINISM : ELIMINATION MADE ABSOLUTE—MR. A. SUTHERLAND

A strong book with some weaknesses—Works out the origin of moral feeling by natural selection—Restates Drummond-like position as Darwinian (?)—And exemplifies "arrival" of forms—*Biology;* fitness to survive—*And* to breed *and* rear—Quantity first relied on—Then quality—This develops sympathy—Which becomes serviceable—*Anthropology;* everything depends on the approaches to monogamy—*Sociology;* progress is by elimination of the inferior—Even when it *seems* to find more rapid means—(Yet he allows *some* progress by imitation !)—*History;* retrogression is possible !—For he hates all militarism—On the whole he does not believe in history—Or in reason—*Ethics;* Has dealt only with one-half of goodness !—Egoism must balance sympathy !—Balance will grow automatic !—Criticism ; no right to call sympathy *moral*, if only half of morality—Nature does not select one quality at a time !—Selection said to have worked—Not *true* natural selection though—Why is goodness not automatic already ?—Do beauty and goodness exist, or do they not ?—" Yes and no ! "

MR. SUTHERLAND'S two handsome volumes are among the most recent, and certainly not the least important, contributions to the biological study of morals. They are interesting in many ways. As a gift from Australia to older lands they deserve a courteous welcome. As the outcome—so we learn from the preface—of eleven years of labour they deserve our respect and almost our reverence. They cover a very wide field, including biology, anthropology, history, philosophy. In the first Mr. Sutherland gives many results of his own observation, and so far as a non-expert can judge, he seems admirably equipped both as observer and as summariser for speaking on questions of biology. The same might be

said regarding anthropology. In history Mr. Sutherland does not profess to be an original scholar, but he
quotes to good purpose, and generalises strikingly. Yet
why does a student of Robertson Smith express himself
as if he had never heard of Old Testament criticism?
Why should he speak as if the character or conduct of
King Solomon threw any possible light upon the Book
of Proverbs? No doubt the Old Testament references
are of trifling amount; but when an author is dependent (necessarily) on a great amount of borrowed
material, one cannot but judge of his quotations from
regions beyond one's knowledge by what one sees of his
procedure in regions where one is able, so far, to control
his method and test his judgment. In philosophy,
finally, Mr. Sutherland is well read, but is hardly
master of his materials. A writer who supposes that
Kant's "moral law" meant the statute law or criminal
code, puts himself out of court. And, for our part, we
must dissent in the gravest possible way from his
philosophical principles.

Mr. Sutherland is chiefly interesting to us from the
unflinching way in which he carries out the appeal to
natural selection, or, as he very tellingly words it, to the
working of "elimination,"[1] in one region after another.
He conducts a valuable experiment in seeking to use
this one conception as a key to all the mysteries of
progress. Mr. Sutherland modestly tells us that he has
done little more than expand Darwin's chapter in the
Descent of Man. Yet Darwin was concerned with
morals only in an incidental fashion. Morality furnished
a possible objection to the opinion that man is descended

---

[1] Yet it is questionable whether Mr. Sutherland's elimination is the
same process throughout as Darwin's, *i.e.* whether his natural selection
in morals, etc., is true natural selection.

from brute races.   Darwin rebutted the objection by
showing the affinities between human morality and
animal sociality.   He did not trace out in detail the
derivation of the one from the other by the working of
natural selection ; and this Mr. Sutherland does, or seeks
to do.   The appeal is steadily made to natural selection,
and natural selection alone.   Use-inheritance is "a
matter under discussion, and on the whole improbable." [1]
Reason is in no sense conceived as modifying the workings of selection which we see in nature.

A second feature of special interest in Mr. Sutherland's book is his ingenious restatement of views very
like Henry Drummond's in the *Ascent of Man*, and
his restatement of them as the legitimate outcome of
the Darwinian tradition.[2]   To at least one reader Mr.
Sutherland's account of the animal anticipations of
morality has made the point of view intelligible and
impressive as it never was before.   One cannot doubt
that there is a rehearsal of the whole drama of morals
in races lower than man.   And one learns from Mr.
Sutherland how sympathy, which he treats as the
primary form of morality, was actually a factor in
securing further progress.

Yet a third reason for valuing Mr. Sutherland's book
lies in the instances it points out of progress coming to
its limit in certain directions, and so terminating.

We must now try to describe briefly the leading
thoughts of this full and interesting discussion with its
admirable wealth of examples.   We begin with biology.
The first of all necessities is that emphasised by

[1] ii. p. 89.
[2] Yet this is rather a transformed Darwinism.   It gives a more *moral*
view of the animal world (not of the human !).

Darwin's doctrine, that the individual organisms should be fit or fittest to survive in the endless struggle of life. This postulate, however, does not carry us very far. The individual may survive, but the race will not survive or preponderate unless the victorious adult organism is able to bequeath its position to offspring, and thus to reproduce its great qualities—the congenital, if not the acquired qualities — in a subsequent generation. Of course the converse is equally true. There can be no transmission of qualities unless there is first, and for a time, personal survival! Therefore, Darwin's postulate may occupy the first place in our list of requisites. But the course of discussion has made the position clearer. It is not individual organism that competes against individual, but stock against stock. The prize of survival goes not simply to individual strength, but to individual strength *plus* an abundant healthy offspring.

Now there are two ways in which nature has secured, and does secure, the maintenance of species. One is the method of quantity, the other of quality. In the lower forms of life, and in some which are pretty high, fecundity is almost inconceivably great. But the superior method is that of quality. Fewer of the offspring perish at an immature stage, for they are better guarded and better developed either before birth or while still under parental care. The methods are alternatives. As quality rises, quantity recedes. As care for offspring increases, the number of offspring steadily diminishes ;[1] but every species pretty well holds its own on a net balance. One important side development of the method of quality is the method of the egg, the nest, and the

---

[1] Does this not point to a variation which is not *random?* Are we really to suppose that, in the beginning, animal races produced families of all sizes, indiscriminately, and tended them with all possible degrees of care, until those with unsuitable proportions died off ?

incubating parent; but the crowning method is that
of infant helplessness and maternal or parental self-
sacrifice, best exemplified in human kind.

We see therefore that the higher races are evolved
on a principle of family life and family affection. But
in this close intercourse of the home or the nest sympathy
is born, and sympathy naturally extends itself to other
members of the species. Here then we are on the very
brink of morality itself. Indeed, we might say that
the secret of the evolution of morals is placed by Mr.
Sutherland just here. Nature, in the case of the higher
tribes, required for survival that there should be a strong
" perihestic " sympathy, and this sympathy could not
be hindered from overflowing into " aphestic "[1] relations. Morality was, so far, a kind of by-product in
evolution, though an inevitable by-product. Family
sympathy was a necessary cause of predominance in
those races which had substituted quality for quantity,
care or development of offspring for mere fecundity;
but in the first instance germinal morality, or the wider
sympathy, was a symptom rather than a condition of
progress.

Only, however, in the first instance ; for as animal
life drew nearer and nearer the confines of morality,
and even before it had grown rational, gregariousness
or sociality became serviceable.[2] The more gregarious
were selected, the less social were eliminated.

Here then we have Drummondism brought into
relation with natural selection, and exhibited as a sub-
section in the Darwinian theory.

---

[1] Mr. Sutherland's terms, coined by him for human morals, where no
doubt they are more fully legitimate.

[2] Or so it is argued. The shoal darted away when one fish saw
danger ; yes, but did not the shoal become a mark for dangers which
solitary individuals might have escaped ?

In anthropology Mr. Sutherland is inclined throughout to emphasise the importance of monogamy, and of the poorest, most imperfect approaches to it—never conceding much sway to polygamy, and not attaching importance to those strange phases of social development studied, *e.g.*, in connection with totemism. In other words, Mr. Sutherland—like Mr. Herbert Spencer, though in different form—holds that there were no very complex processes involved in making man so social as he is. It is natural that such views should be advanced by one who puts the centre of moral development in the family, and who believes that all development—moral development, infra-moral development, development of morals out of the non-moral—is due to natural selection. Mr. Sutherland's views are supported by much evidence as to the character of contemporary savage life. But, if other reports can be trusted, there are features both of the present and of the past which deserve more prominence than they receive with Mr. Sutherland.

In general sociological theory Mr. Sutherland is strikingly loyal to his doctrine of elimination. Human or moral progress is due to elimination, not by means of wholesale massacre, but through the gradual and unnoticed working of natural law. Criminals as a class leave but few children ; necessarily therefore, in a generation or two, criminal stocks die out [1]—or, shall we say, *tend* to die out ? The vicious and grossly self-indulgent produce or rear few children ; they also die out. Even the coarse and violent tend to kill each other off. " They that take the sword perish with the sword."

[1] What about the Jukes family ? And again, if a criminal population is generated afresh by society at each stage, *have* we advanced by the elimination of previous criminals ?

The meek inherit the earth by the simple process of
"lyin' low and sayin' nuffin'," like Brer Fox, or like
the Babes in the Wood, while the ruffians dispose of
one another. All this is vastly well so far as it is true;
but the violent, at any rate, have no special taste for
singling out their violent rivals; they are quite as ready
to murder, outrage, or plunder the most sympathetic
and inoffensive of their neighbours.

Let us observe however the full force of the position.
This method of elimination is regarded as *the* method of
moral progress. It is so certain and so telling that all
others may safely be neglected. When Christianity
was accepted by the Teutonic barbarians it did not in
the least pull them up to its higher moral level. Slowly,
in the course of some thousand years or so, the incapable
were weeded out and the general level was raised. Those
sinners wandered in the wilderness for very nearly forty
generations till the whole stock died out in detail. This
is a doctrine of the most unbounded materialism. It
regards man as fatally determined by his antecedents.
Free-will is a dream, conversion or real repentance an
impossibility. Yes, and that is all implied in the attempt
to run natural selection right through—to make elimination the only method of moral progress.

At one point Mr. Sutherland seems inconsistent with
himself. In one passage he almost bursts the shackles
of naturalism. He speaks of imitation as a cause of
progress—like Bagehot, or like Professor Baldwin. But,
so far as imitation acts, elimination is unnecessary. If
example can be copied, there is a short cut to progress
on the part of the inferior but teachable multitude. In
nature imitation plays a very limited part. One species
cannot borrow the good habits of another. If it could,
you would have transformations ready made without the

cumbrous machinery of elimination. And if imitation does work in human history, then, so far as it works, it supersedes natural selection.

We may make a separate heading for Mr. Sutherland's conception of history in detail. The method of elimination being always steadily and triumphantly at work, we seem to have before us a programme of the boldest evolutionary optimism. All must be for the best in this best of all possible universes. Progress, it would seem, cannot fail or be checked. That, we think, ought to have been Mr. Sutherland's doctrine, given his premises. Yet it turns out that he believes the clock went back precisely one thousand years when the barbarians overran the Roman empire. It took the barbarians precisely that time—Christianity and all— to reach the social and moral level of ancient Rome (!!) —and then progress recommenced. Now, what does this singular view mean? Perhaps for one thing it means that Mr. Sutherland—like Mr. Spencer, yet not altogether like him; unlike Bagehot—has no sense of the moral worth of war, under whatever circumstances waged. It means that the masculine ideal, in spite of some isolated references to it, is left out of the reckoning, while the feminine ideal of sympathy is given a place of absolute predominance and authority. In a world wholly governed by natural selection, softness surely ought to be ranked as a deadly sin. The Roman empire had grown too soft to fight. It was not therefore advanced, but retrograde, and unfit to survive. The barbarians may have been one thousand years behind, tried by certain tests; but, in the light of the most practical of all tests, they were not behind, but before. Of course Mr. Sutherland's ultimate definition of

" morality," as we shall find, makes it only one constituent of human well-being.   Surely a very unfortunate abuse of terminology in a moral treatise!

Another qualification of Mr. Sutherland's views—by
common sense—slips out when he speaks of Howard
the philanthropist[1] as moving his age.   Now, this is
curious.   Christianity had no chance with the Teutonic
peoples till natural selection killed off the heathen and
barbarous majority; John Howard, without waiting for
natural selection to make "*Howards* of us all," was
able to "*move* the hearts" of his fellow-countrymen.
And yet Howard, with all his qualities, was surely not
comparable to the founder of the Christian faith?   The
one had his *milieu* ready made; the other had to create
his *milieu;* but was His greatness not tolerant of that
extra burden?   Or put it at the lowest : if personal
influence is capable of doing *anything,* is there not a
factor in moral progress to be reckoned with, independently of natural selection?

On the whole, however, we might almost say that
Mr. Sutherland does not believe in any such thing as
history, or the throbbing and thrilling of the social
organism to one great life.   In history the public mind
" moveth altogether if it move at all "; whatever lies
below consciousness, there is a conscious life, and the
conscious service of common ideals.   But Mr. Sutherland
will have it that nothing ever happens, except the inter-
minable weeding of the human garden.   The bad die
out; the good have only to stand still, and they, or
their stock, will be carried on by forces outside of them
to a far-distant triumph.   We are in no sense members
one of another.   We are not so much men as things—
things exactly like other things—or exceptional only in

[1] i. p. 420.

this, that we can find out in what direction we are
tending, while we are utterly incapable of modifying
that direction or of altering the pace.

Along with Mr. Sutherland's doctrine of history we
may take his doctrine of reason, which resembles the
other doctrine closely.   There is no such thing as
reason.   Applying natural selection to every process,
from the life of the amœba to that of the saint, Mr.
Sutherland scarcely has room for reason in his system.
And therefore he shows us nature selecting the fittest
emotions in the form of so many physiological processes
—consciousness being a mere blind alley; it came
no one can say how or why; it leads nowhere.   The
appropriate emotions are organic to our race, in total
independence of the accident of reason or consciousness.
They might last if it lapsed; they are untouched and
unaffected by it.   It is a practical nullity, and ought
not to have troubled our theories by existing at all.[1]

Passing on to morals, we meet with the great surprise
of the book.   By "moral instinct" Mr. Sutherland means
sympathy.   There is, he says, no instinct which tells us
what is right and what is wrong; moral opinion could
not vary as it does if instinct were controlling it.   There
is, however, a sympathetic instinct—the creation of
natural selection.   And that instinct tells us one of the
conditions of right conduct; another of the conditions,

---

[1] Mr. Sutherland ascribes emotions to a bodily source, and remarks
that Professor William James has reached similar views.   One observes,
however, that Professor Lloyd Morgan speaks of the "almost paradoxical
emphasis of Mr. James's views," and of "making them somewhat less
repugnant to common sense" by confining them to the first rise of
emotion, in contrast to subsequent emotions qualified by "association."—
Habit and Instinct, p. 190.   Dr. S. H. Mellone (*Studies in Philosophical
Criticism and Construction*, p. 249) states that Professor Dewey has maintained the paradox with more determination than Professor James.

however, is dictated by the egoistic instinct. Right action is a resultant of these forces, or a compromise between them. Here, then, our great Darwinian in morals suddenly becomes a Spencerian in morals. And he goes all the way with Mr. Spencer. He looks forward to an age of perfected balance, when good conduct will be automatic; when there *will* presumably be a moral instinct! Natural selection, steadily killing off the inferior types, will at last produce that " crowning race."

Now, is this fair? Truly, it is easy to show that morality is an outgrowth of sympathy if you define what is " moral " as equivalent to what is sympathetic! All the time we are reading Mr. Sutherland's record of moral evolution we suppose that we are being shown the gradual origin of real central goodness,—of that spirit which embodies itself in right conduct, and does so willingly. Suddenly we learn that our impression was wrong; that was not what were being shown; we were looking on at the production of one constituent of goodness, but the other constituent, which is no less important, is quite a different thing! Then were the morally advanced Romans, who succumbed before the barbarian inroads, *not* really better men, but just more sympathetic? The whole book had need to be rewritten. Mr. Sutherland must not talk of morality if he has in view only one half of its conditions. Language has its rights, and the truths embodied in language must not be flouted, or they will take their revenge.

Moreover, it seems very doubtful whether Mr. Sutherland is entitled to assume that natural selection has developed sympathy, but has developed it in uncertain measure, so that it may be perhaps too much in amount, perhaps too little. Natural selection has taught the lower and the higher animals *exactly* how many

offspring to produce. Why has it not taught me
exactly how much sympathy I am to feel? Why has
it developed a force uncertain in amount and working?
Unless because, after all, spirit is different from nature;
because it is inconceivable that natural selection, and natural selection alone, has " out of darkness " stretched forth
" the hands that reach through nature, moulding men."

Again, let us note that two qualities have been
selected—two seemingly if not really opposite qualities
—egoism and altruism.[1] How may this be? Nature
has really been selecting *men*, not qualities, men (or
societies of men) who are the sum of all their qualities.
Nature is regarded as a hanging judge. Every crime
in her calendar is a capital offence. If nature is not
satisfied with you, " Off with his head," she cries; and
forthwith you are thrust out. Nature has not been
selecting one quality at a time; she has been selecting
aggregate fitness. It is lawful to study the process one
quality at a time, if you like. But you must keep in
mind that that is your own " abstraction." The only
question with nature has been, first and last, who is in
the aggregate fittest to survive? Fitness has been
selected, not quality A tending to fitness, nor quality B
tending to fitness, but $A + B + \ldots M$. And again,
from this rather different point of view, we are struck
with the anomalousness of the fact that natural selection, if it has really been at work, has not already
produced an automatic balance between egoism and
altruism, or has not done so in the past if it is going to
do so in the future.

There might indeed be an explanation, if sympathy in its wider range outside the family were only
(what Mr. Sutherland holds it was primarily) a by-

[1] Mr. Sutherland thinks the latter word stilted, and avoids it.

product in evolution. In that case sympathy ought to be a casual and fluctuating factor in human nature.[1] But Mr. Sutherland carefully rules out that view. Sympathy has been in the main a condition of success, and has been selected as such through untold ages. Is not Darwinism, at least apart from statistical tables,[2] a dangerously plastic method? Anything and everything may be conceived as a quality tending *in some way* and *to an undefined degree* towards predominance. Anything and everything may be ticketed, "First prize, for fitness to survive." The formula of Darwinism

> Is twice too big,
> And therefore needs must fit.

Indeed, one observes that, in spite of his Darwinian phraseology, Mr. Sutherland is not thinking of natural selection *per se* as an evolutionary force, but of natural selection modified by the presence of animal sympathy. This seems a true account of the facts of nature, but it is a miserably inadequate account of the facts of human society; and unfortunately Mr. Sutherland admits no morality among men beyond the rudimentary morality which he finds in the brute world. Elimination must do everything for us; it cannot! And whatever elimination does for human advance it is precisely *that elimination which is least like the Darwinian* that survives the advent of reason. If the child of vicious or criminal or heartless parents is neglected and dies, while the child of honest, pure, and affectionate parents sur-

---

[1] Compare Mr. A. J. Balfour's remarks upon the æsthetic sense (*Foundations of Belief*, Book IV.), based on the assumption of evolution by natural selection.

[2] Demanded by Mr. Karl Pearson in *The Chances of Death, etc.*—Dr. Pearson, one notes, is a Professor of Applied Mathematics.—His suggestion deserves consideration.

vives, there is no struggle. The better care paid to the
second child is not the cause why the first succumbs.
If the ill-cared-for child were the only child in the
world, it must still die of neglect. "Elimination" here
is not a case of selection after struggle ; it is nature's
own protest against vice and exuberant selfishness.

But let us pursue the subject further. Does Mr.
Sutherland habitually place himself outside of morality,
and view it with scientific coolness, as one quality
tending towards success? Or does he write from the
inside, with a glow of admiration for "the true, the
just"? Very often he does the latter. It would be
altogether to misrepresent Mr. Sutherland if we did not
confess that he writes like a good man and a lover of
goodness. But in his final attitude he seeks to combine
both views. Goodness is authoritative for us ; we are
bound to be loyal to it ; we must speak and think and
feel as if goodness were something objective and absolute,
cosmical, divine ; and yet reason forces us to be agnostics.
Goodness is nothing but one of the conditions of race
efficiency and race survival. Beauty is nothing in
itself ; and the sense of beauty is mere habituation to
environment, whether from inherited experience (Lamarck, Spencer, also Darwin) or from the slower but not
less sure (Darwinian) process of elimination. We must
steadily occupy a position on *both* sides of the hedge.
Mr. Sutherland is determined to warm his hands, as long
as he lives, at a painted fire. He knows it is painted ;
you shall not throw dust in his eyes ! He is determined
to keep on warming himself ; how dare you forbid him ?

We at least have no wish to do so. We would rather
hope that some day he may discover the glorious truth,
that what warms him is not paint, but God's own
sunshine.

CHAPTER XVII

THE METAPHYSICS OF NATURAL SELECTION

I. Chance in relation to purpose, as *accident*—As *absence of design*—In relation
to law ; as *blind* law—As *blind combination of laws*—Compare with the
last the scientific or mechanical view of the world ; a number of separate
substances ruled by a number of independent laws—Good enough for
science, not for philosophy—Darwin ought not to assume things as *really*
disconnected, merely because he has not *needed to investigate* their connection—As if organism and environment were accidentally brought together—
Or as if organism and organism were *mere* rivals—(They *are* rivals !)—Or
as if force and force were disconnected ?

II. Darwin treats *variation* as casual, *i.e.* as a thing with no bearing in itself on
the purpose of the species—His theory *allows* this assumption—But does
not *prove* it—We all habitually understand the theory in that sense, *e.g.* in
contrasting natural selection with use-inheritance—On the fact, evidence is
wanted—Conceivably variation may choose very irregularly between many
fixed possibilities—This seems to point back to disconnected laws, as in
last section.

III. Even on Darwin's own view he is hardly entitled to call the process of
evolution natural selection—Aggregate range of possible variation is fixed
by the nature of the material—Two agencies must be taken together—Of
the two the varying organism, not the blindly selecting environment, seems
the better to account for rise of new qualities—Summary of I. II. III.

IV. Kinds of natural selection, A, B, and C—B exists !—If organic evolution
is a fact, C exists !—Accelerating any other evolutionary force that may
exist, and of course involving B—If A is found *alongside* of C, A must have
a separate field where C cannot enter, else inconsiderable—Natural selection
(C) lasts as long as nature is nature—Even along with (the more rapid
force of) animal intelligence—True reason checks it—Does natural selection
ever work by itself (A) ?—Higher animals with fewer births evolve as
quickly as lower ; has a new force arisen ? or was natural selection never
the leading force ?—[Can we regard *intelligence* as the new evolving force ?
Dr. Mellone assumes its operation everywhere !]

V. Can natural selection apply to men ?—*Biologically*—Struggle with beasts is
over—Famine (A) is rare, and of doubtful tendency—Pestilence (C) does
harm—Vice (B)—Crime (B)—War (selects the wrong way)—Religious
celibacy (*ib.*)—Summary—*Sociologically*—Mr. Kidd's insistence on struggle
is really biological ; is unproved ; is not an insistence on natural selection
—*Ethically*—Mr. Alexander's competition of "*Ideals*" is exaggerated—
And itself implies reason and sympathy—Mr. Sutherland's elimination of
evil *doers* ignores positive causes of moral progress—Exemplified typically
in Jesus Christ.

VI. If natural selection does not operate where reason and conscience exist, it
   yet may *originate* them in the loose and incorrect sense in which natural
   selection is said to originate things !—If reason, etc., were, as most suppose,
   evolved and selected—How selected ?—Have *adjacent races* died out ?
VII. Other idealist views—Professor Ritchie praises natural selection more fully,
   in vague terms and in some passages—Mr. Sandeman rejects it, because he
   believes in the teleological perfection of every organism—But is it possible
   to get over the impression produced by rudimentary organs ?—It is enough
   if the *whole* of nature is *good*, and its *parts relatively fit*—Dr. Stirling
   believes the casual variation which makes an individual can never make a
   type—Is it certain that every individual is born differentiated ?—Or that
   any differences are incapable of growing by cumulation into a type ?—
   Possible value of the hypothesis of natural selection, even if a fiction.

IT was no part of the plan of this book to undertake a
direct criticism of theories of evolution upon their
merits, whether from the point of view of biology or of
philosophy, of science or of metaphysics. If we now
find it necessary to undertake an estimate of the value
of Darwinism, we do so not merely because of the outstanding importance of that theory, but because, in
summing up results, we are led to insist on a distinction. While we admit, and even (so far as we have any
right to speak) defend, the theory of natural selection in
biology, we affirm that it cannot be applied in sociology
or morals. Such a view seems to need justification. It
can only be supported by a review, however hurried and
imperfect, of the merits of Darwinism.

The question may perhaps best be approached by a
discussion of the element of *chance* contained, or said to
be contained, in the Darwinian theory. Perhaps some
minds love Darwinism, because it appeals to chance ;
others undoubtedly distrust and despise it for that
reason. What is chance ? Does Darwinism assert
chance, and, if so, in what sense ? How far is it warranted in doing so ?

I

First and most simply, chance is the opposite of
purpose. It implies a failure of purpose where the

presence of purpose and its successful realisation were expected. A train is meant to carry me safely to my journey's end—that is purpose. Instead of doing so it runs off the rails; the natural forces set to work were imperfectly known or imperfectly controlled. That is accident, not purpose. Neither the passengers nor the company's servants designed that result. When a young rough puts a stone upon the track, and wrecks a train, that is not "accident," though by a natural extension of the term we may call it so. That is not chance, but wicked purpose. It is crime.

Darwinism does not exactly assert chance in this sense, although it may seem to do so. Apparently Darwin himself believed that he had destroyed the evidence in support of purpose or design in nature. J. S. Mill too, looking at the new doctrine, thought that, if it were established, it would substitute chance for design. The evidence for the latter would go to pieces on the "plurality of causes." But even if Darwinism should be held to destroy teleology, such a view involves using the word "chance" in a sense markedly different from that in which we have defined it above. Chance or "accident" in human life means partial failure of purpose through man's weakness or ignorance—partial failure standing out in sharp relief against a background of habitual success. He aimed, as he always does, but he missed the mark this time. That is what we mean (so far) when we say "the disaster was due to chance"; "he had a dreadful accident yesterday." There is no full parallel between this and Darwin's wholesale denial of teleology in nature. *There was no one to take aim,* hints Darwin.

Moreover, it is not enough to deny teleology. It is necessary, if you are to carry weight, that you give a plausible explanation of the fact that nature mimics purpose. Darwin has given such an explanation. What part does chance play in it?

If we cannot fully interpret chance by a reference to telic purpose, we must bring it into relation with efficient causation—or causal law, as we ordinarily phrase it; efficient cause, or that scientific conception of cause which stands nearer to efficiency than to any other of Aristotle's "causes," having well-nigh monopolised the name of cause in the minds of modern men.

The assertion of chance will now imply either (1) mere blind causal law, in opposition to purpose, or else (2) mere blind coincidence of several unconnected laws or forces.

The phrase is often used in the first sense in denunciation of Materialism. Did mere blind causal law, it is asked,—did the mere law of matter blunder into *mind?* This, however, could not be Darwin's sense. He denied purpose; but it was not at all his affair to disparage causal law. Besides, it is not the case that any one cause (not being a mental "First Cause") can be said to account for living species. "Natural selection," the supposed creator of distinct "species," is a group of many different causal factors, curiously entangled with each other.

We are driven then upon the last sense. A chance is a *coincidence*. Series A and Series B cross each other at one point, and affect each other unexpectedly—it may be, grievously. They are distinct things; but they "happen" to have their existence side by side in the same universe; presently they "happen" to exchange their formation side by side for a hostile formation, front against front, and there is a collision,—it was an accident! The wind that blew over the rotten tree, the cry that caused the child to run forward, had no connection with each other. But the child "happened" to be just under the tree as it fell, and was crushed—by accident.

The champion of ethics must not look askance upon the doctrine of chance in *this* sense. Chance and choice are very closely connected. Man can neither create nor annul force. He can govern it only by determining *where* one current shall cross another. No contingency in nature, would imply, No free-will in man! or at least, no power of affecting external nature by his will.

Moreover, this form of the doctrine of chance, or something very like it, is involved in the logic of science. We call it *mechanism*. The "finite" sciences take a *mechanical* view of the universe. They reduce its processes to a few elementary substances (chemical elements, *e.g.*), actuated by a few elementary forces. Sometimes, as in Mr. Herbert Spencer, we find more fundamental views of evolution proceeding spontaneously from a homogeneous material unity; but such views are a dreamy speculation; they have neither the demonstrativeness nor the definiteness which are the glory of science.[1] Science is content to pause—where perhaps it thinks that knowledge itself pauses—at the discovery of distinct separate substances and distinct separate forces. And so to it the universe is a machine—not an organism; the co-operation of distinct parts explains

---

[1] This characterisation may seem to ignore the law of the correlation of forces or transmutation of energy. But how far does that law carry us? What does it affirm? Different forces are different manifestations of one force, taking their shape *under different given conditions*. I do not see that science can simplify beyond that statement. Accordingly, the given conditions represent the "ultimate" plurality, with which scientific analysis leaves us.

the cosmos; its unity is not (as in an organism) prior to the distinction of parts from each other. May we take it that, as long as we are thinking in terms of matter, this view is correct? That such a mechanical view of the universe is the ideal goal of (finite) science? Speculative thinkers will ask for more. The mind itself may demand some deeper or fuller unity. Are not the different substances in some way calculated or adjusted or related to each other? Is their coexistence purely casual? Is the quantity of each (so far as we can speak of quantity in the whole universe—so far as we can treat the universe as finite) purely casual, or is it determined by some obscure law? These questions lie beyond the range of the special sciences, which carry on their business quite successfully apart from such researches, finishing their own work upon the crude assumptions of mechanism—a few substances; arbitrarily given quantities of each; a few elementary laws. Possibly, as we have said, you cannot reasonably go further unless you quit the logic of science for philosophy — unless you exchange matter for some frankly idealist conception of reality.

Within science, then, there seems to be a doctrine of coexistence closely analogous to what we mean in ordinary speech by chance. It differs in one respect; "chances" are occasional interferences, while science details the habitual co-operation of law with law. The difference supplies science with one excuse for declining to endorse an appeal to mere "chance" on the part of Darwinism. But the conceptions of scientific mechanism and of chance coexistence are identical at heart. Both take as given several independent substances and processes, without asserting or believing in any wider law connecting them with each other.

There is indeed a different way of escape besides the metaphysical shifting of the point of view. We may address ourselves to old‑fashioned teleology. Keeping the idea of hard, repellent, individual things, we may suppose that a designing and combining force, external to themselves, crushes them together into a unity. But such a philosophy is liable to be charged with dualism. And not without reason; it is quite as mechanical, in its own way, as the logic of science. Here once more two elements, which as the Germans say "belong together," are made to fall asunder. The material elements or forces, and the law of their combination, are assigned to different quarters. Nature has no tendency in itself towards life; a Deistic God outside of nature forces *His* thought of life upon alien materials, as the human sculptor forces the design of his brain upon the marble, which was fused in nature's laboratory without any reference to the needs of artist or artisan. Hence also it is clear why a system of idealism, which tries to show that all things are related together, and especially that design and materials belong to each other, becomes suspected of pantheism. There is undoubtedly a pantheistic strain in it. Are we sure that there is not a pantheistic strain in the truth and nature of things?

It is not any form of teleology, but, on the contrary, the purely and characteristically analytic procedure of science, that we seem to find in Darwin. With him, natural selection is a biological hypothesis. He proposes to account for all the different living species from a few given elements—(1) organisms, multitudinous in number but simple in kind, distinct from each other, hostile, competing for the prize of survival; (2) an environment in which life is possible;

(3) heredity; (4) variability. The first three factors
sum up in the result (a) struggle; and all four
factors taken together give us the final result (b)
selection. The immediate outcome of Darwin's theory
as a contribution to science is this, he needs *no
additional factor.* The factors already named suffice
(he holds) to account for the further result—many
distinct living species. As a scientific worker, Darwin
simply postulates his small array of causes existing
casually alongside of each other. The man of science
has no need to search more deeply, and Darwin does
not do so. But, when natural selection is generalised
as a philosophical theory, when it is applied to other
departments of existence, outside of and above biology,
we must raise deeper issues. We must not allow the
assumption to pass as matter of course, that the
"abstractions," which are legitimate and necessary in
special sciences, are facts, or are determining conditions of all human thinking. Because you have
skilfully dissected the world into a few separate limbs
or tissues, and can show exactly how they fit together,
it does not follow that no subtle "spiritual bond"
has eluded your scalpel. Because you can explain
your special problem without asking whether organism
and environment, organism and organism, force and
force, have any necessary relation to each other beyond
the bare fact of coexistence, it does not follow that
you have demonstrated the unreality of such a relationship. You have assumed its non-existence—
or rather you have ignored the whole question; and
quite fairly for a special purpose. But you have proved
nothing. And the sceptical programme is improbable,
perhaps impossible ; "mechanism made absolute ; chance
the only nexus between the elements of nature!"

Such is the view of Darwinism which I suggest. Those who entirely reject natural selection, even as a biological hypothesis, may insist with a good deal of force that organic life—that curious half-way house between nature and spirit—or may insist that animal life, so far as psychical, already shares largely the nature of spirit; that therefore we are guilty of folly in treating it on physical or mechanical lines. If in an organism the whole is prior to the parts, can we explain the genesis of organic species by the coexistence and interaction of [things which we treat as] distinct parts? The objection is forcible. Does it not amount to saying that a *science* of biology is impossible? That philosophy must annex to its own department all treatment of the problems of life? I think such a view extreme.

Let us see how the doctrine of chance or of mechanism works out in sundry particulars of the Darwinian hypothesis.

Organism and Environment. — Darwin assumes elementary living forms (else he has nothing to make species out of), and plenty of them (else there will be no struggle). He takes them for granted: they have a suitable environment; they live and are able (some of them) to survive. It is not his affair to ask whether organism and environment have any mystic connection. He takes them as given. They are facts—just facts.

Yes; but it is a very long step indeed from this point of view to the denial of teleology, to the assertion that organic fitness itself arose through natural selection by the weeding out of unfit forms. The ignoring of the problem of necessary relation between organism and environment is one thing, the denial of such relation is quite a different thing, and nothing in scientific

Darwinism justifies it.   Darwin the biologist has shown
us how life may advance, build itself up, differentiate
itself; how fit may become fitter.   He has not shown
us how unfit may tumble into fitness.   Among the
postulates of his process of biological evolution are
numerous fit living forms.

Organism and Organism.—These, Darwin tells us,
have nothing to do with each other except to struggle
against each other.   Not all creatures stand directly in
relations of struggle.   Probably a whale and a robin red-
breast have no influence on each other's estate.   But,
when organisms do affect one another, they do so on
terms of hostility.   Some species prey upon others.
In adjacent species, and within the same species, there
is (from our point of view, not from theirs; they have
not consciousness to intensify it), there is competition
for nourishment.   All of them cannot survive times of
scarcity or danger.   The weak have their chance but
get weeded out.

This statement ignores (1) animal sociability and
mutual help, usually, not always, between creatures of
the same species.   Competition, it may be argued, is
largely a human surmise or interpretation; sociality is
a fact, psychical as well as physical, in animal life.
(2) It ignores the *dependence* of animals upon living
food of some kind.   True, the relation of the eaten to
the eater is not one of friendship.   Yet it is a highly
positive relation.   It is not the whole truth about the
cosmos of life that its many species and innumerable
organisms are inconsistent with each other.   The food
species does not simply struggle against the predatory
species by flight, mimicry, protective organs, etc. ; it
makes such species possible.

Having made these deductions from its value, can

we accept natural selection by struggle as a (or *the*) great method of evolution and lever of progress in nature ? There is no great presumption, surely, in putting the question ! The evidence in favour, not of organic evolution, but of natural selection as its method, is deductive and hypothetical ; the same thing indeed is true of many of our scientific theories. The evidence for natural selection is as follows : (1) Struggle and selection are facts ; (2) They will—given time enough—account for quite as much progress, quite as much differentiation as we see in the cosmos of life ; (3) Therefore, by the law of parsimony, they have caused it. All this is only probable evidence, and "the plurality of causes" may undermine it. Accordingly we claim the right of criticising the theory, and of asking whether it is antecedently credible. Is it thinkable that the evolution of life proceeds along lines of struggle ?— Surely that is thinkable. The doctrine merely implies that living organisms are parts of nature and are treated as such ; that though the organic and the animal may approach the spiritual, they have not yet reached it. And, by naming one intelligible and thinkable process of evolution in organisms, Darwin has even helped the cause of sound philosophy and the cause of faith. When we meet with intelligible processes, we perceive the presence of reason in the world ; and when the Christian perceives reason at work, he is more than ever assured that the world he lives in is God's world.

Force and Force.—Symmetry with what has gone before would lead us to head our next paragraph with these words. But it is questionable whether we can fairly charge Darwin with treating the different biological forces involved in natural selection—life, variability, heredity — as mutually independent and merely coin-

cident things. Scientific logic may incline students of
science to do this, but a wholesome sense of biological
realities will keep them in check. Where Darwin is
open to question in this region is in his doctrine of
variability. Is variation related in any intelligible
fashion to heredity? Or is it purely "casual"? Perhaps we shall find that Darwin emphasises the mechanical
blending of distinct heredities — that "heredity and
heredity" are pitted against each other in his thinking,
quite in the spirit of the logic of chance.

The question is so important, and at the same time
so complex and obscure, that we had better make a
fresh heading for it.

II

We have to ask then whether there is a special appeal
to chance by Darwin in his doctrine of variations?

Darwin largely treated these as casual, almost as if
uncaused. But it was not, for the moment, his affair to
say how variations arose; he was to show how they
worked out. He never thought of asserting deliberately
that variations are uncaused; his followers explicitly
deny and repudiate any such view.

What Darwin has done is to assume that variations
are casual in reference to the purpose of the species;
that the individual variations arising in nature, so long
as they are unweeded by struggle, do not directly tend
to fitness. In this sense Darwin affirms, or rather
implies, chance—chance in contrast with purpose, but
yet with a distinct shade of meaning from either of the
senses of chance as against purpose which we noted
above. Not (1) partial failure of purpose is implied,
as when men fall into accidents. Nor yet (2) entire
absence of (proved) purpose, as when Darwinism is said

to destroy the teleological argument for the being of God. But (3) partial absence of purpose. While all the other processes of plant or animal life are purposeful, variation moves at random.

Darwin we say assumed this. He did so when he called the entire process *Natural Selection*. If variation itself were (to any extent) purposeful, progress would not depend entirely upon the selecting agency; but Darwin's nomenclature implies that indirect selection is the only cause of progress. He had invented a theory which would account for evolution even if variations were non-purposeful. It was natural to slip into a habit of speaking as if variations had been proved to be non-purposeful. But that had not been proved. Nothing had been proved about variations. And so long as we are without laws of variation, it is very hard to define the meaning and bearing of Darwinism.

For example, the general bearing of use-inheritance is naturally defined thus : it will give the same results with natural selection, only *more rapidly*. But in speaking so one assumes, what is habitually assumed, and never proved, that variation is casual, *i.e.* non-advantageous (in itself and on the average). If it turned out that variation moved even in part along the lines of evolutionary change, then Darwinism or even Hyper-Darwinism might warrant the hope of rapid progress. Hence it is extraordinarily difficult to bring to the test of experiment the questions between the Lamarckians and the Weismann school. One glides into the habit of thinking that it is mainly a question of pace. And yet quick pace, if it were proved, might not be a presumption in favour of Lamarckian use-inheritance. It might only point to a neglected element

in Darwinism, to the necessity of regarding variation
per se as telic not casual.

We do not mean here to affirm that variation must
be advantageous, or even that it must proceed along
definite lines. We merely claim that such possibilities
should not be forgotten. The questions are questions
of fact, and further evidence is required. Causeless
variations are inconceivable things; in that view, presumably, all will agree. But, just as little as the
evolutionist would waste time over a hypothesis which
involved surrendering the causal law, so little would
others consent to trifle with a great question by framing
the hypothesis of variations perversely opposed to the
specific type. Still, within limits, we might conceive
of "casual" variations, if variability worked along one
of several fixed possible directions, while the reasons
why it chose one track rather than another were highly
obscure.

Let us take an illustration. Every house of two or
more storeys must include a staircase. The stair may
be straight as a ladder, or it may be spiral, or it may be
a series of straight flights with landings, or it may even
be attached to the outside of the house like the " bonnie,
bonnie outside stairs" at Thrums. The one thing
illegitimate is to omit the stairs, as the amateur who
draws his own plans is so apt to do. Well then, in
variation, the spiral staircase may be beaten into flat
sections, or the outside stair may be brought within
doors, or *vice versa.* Variation may be "casual" in
this sense, that it is liable to take any one of several
directions. Pattern A or B replaces C,—you cannot say
why. Variation will not be casual in the sense of
omitting what is advantageous or necessary. It will
not leave out the staircase. Experience shows that

when animal "monstrosities" occur, they are not strictly congenital. They are the result of accident after development had begun.

As to the reason why variation goes thus or thus in so irregular a fashion ; in a different region one would be inclined to interpret irregularity as meaning the (casual or intermittent) blending of several (distinct) laws, the imposing of several curves one upon another. And so we should be brought back to a "chance" [under obscure temporary conditions ?] blending of distinct influences [parental, ancestral ?].

Tentatively then we would decide that Darwin appeals to chance and that he is right in doing so. He appeals to chance by the assumption that variation is or may be random in its direction,—harmful quite as often as helpful. And — still more tentatively — we propose to identify "chance" in this sense with "chance" in a sense already discussed—the mechanical addition to each other of separate forces interfering with one another's drift. In the present instance, the forces in question are of the nature of hereditary tendencies. But, while we suggest this view tentatively, as good science, we are sure that it cannot be the final truth on the point. The last word upon most topics must be spoken not by science but by philosophy.

III

The phrase Natural Selection.—Thirdly, we have still farther to inquire whether, even on Darwin's own view of evolution, the name natural selection is quite a fair description of the evolutionary process. Darwin the biologist may be right in his facts and causes, and yet Darwin the philosopher may be wrong in the emphasis he throws upon different features in his system, or in the wider suggestions that grow out of his statements of biological doctrine. Now, Darwin's language seems to attribute greater scope to chance than is allowed to it by the deliberate processes of his thinking. The name natural selection seems to imply that progress is due, though negatively and indirectly, to the environment alone. Organisms evolve, it would seem, because of a foreign influence, forcing advance on the reluctant materials. The whole cause of progress lies in the selecting environment, not in the varying organism ; and selection proceeds blindly by destruction of the unfit. Here again we have the spirit of the doctrine of chance. We see it partly in the assumption that organism and environment have nothing to do with each other, partly in the assertion that (if not the existence of life ; to take the same view on that point involves a further stretch of the spirit of materialism ; yet) all advances in life are due to conditions resident in the environment, operating outside and apart from the purposeful processes of the living creature. To say that "natural selection" causes this or that is almost equivalent to saying that " casual coexistence" creates this or that. One is tempted to take up the very opposite position, and assign whatever is new in evolution, even according to Darwin's own analysis, to the varying organism, and not to the selecting environment. "Natural selection" seems a fair enough name for the evolutionary process (as conceived by Darwin), so far as that to which it applies can be regarded as one thing evolving continuously throughout the process. Thus life may be said to differentiate itself into new and finer forms " by natural selection." But natural selection can do no more. It cannot "explain" how matter should pass into life, or how animality should evolve rationality. If for any purpose, or from any point of view, we have to emphasise novelty as novel, then it is unreasonable to speak of the evolutionary process which led to it, even if Darwin's analysis of that process be accepted, by the name of "natural selection." There must have been possibilities in "protoplasm" answering to all the novel results of late evolution. Let the variations come up as they may; let them point in every direction by turns, quite at random, if you insist upon it; still, apart from the amount or direction of each individual congenital variation, there must be a *total possible range of variations*, prescribed *by the material*, and at the very most merely *elicited* by natural selection. Of the two then, life, not environment, the living creature itself, and not the non-living conditions round about it, explains the acquisition of new qualities and the development of fresh specific types. Of the two, Darwin has emphasised the wrong one, and has isolated it by assuming its merely casual relation to the other. So we might speak, in one-sided opposition to Darwin's graver one-sidedness. But the truly reasonable view to hold is that both together—varying organism and selecting environment—and both as elements in one orderly process, lead to evolution.

We do not blame Darwin for speaking in contractions. By the necessity of the case human language is elliptical. The one exception, proving the rule, is furnished by the lawyers. They omit nothing; they recite everything in detail over and over again; and they are the awful example of verbosity, the drunken helots of human speech. But elliptical nomenclature, however necessary, is full of dangers. If I were driving pigs to market I might reasonably (though elliptically) say that

they got there because I headed them off at all the
wrong roads which we had to pass.   Yet it would be
perilous to affirm that "heading off" was the one cause
why they got to market.   They got there because they
were quadrupeds, and disliked being hit.   (I waive, as
possibly not directly relevant, the farther consideration
that there was some one to drive them.)   Yet our
modern evolutionists talk as if barricading the wrong
roads not only kept pigs from straying, but actually
taught them for the first time how to walk.

When we turn to use-inheritance once more we see
that it also may be so developed as to convey the same
vicious suggestion.   New qualities come from without,
not from within ; from the environment, and not from
the organism.   The environment stamps them on the
passive organism, and it (according to the doctrine of
use-inheritance) transmits them to offspring.   But Mr.
Sandeman has forestalled this opinion by a remark of
brilliant force and point.   Every acquired quality, he
observes, is congenital [in its rudiments], and every
congenital quality is also acquired [i.e. developed in
the course of life].   Of course this is a very strong form
of statement, and it seems to forbid all use of the
wonted distinction.   But presently, having fired off his
epigram, and having bowled over his enemy with it,
Mr. Sandeman descends to a less rarefied atmosphere,
and admits that the two possibilities may be contrasted
as matters of fact and [conceivably, though experiment
is difficult here] of evidence.   For the truth is that
every living creature is more or less plastic in definite
directions ; and life develops this *or that;* so it is a
fair question whether or not the offspring resemble the
parent as modified in his own development prior to
his begetting offspring.   But Mr. Sandeman's paradox

serves as a warning. We must not go to use-inheritance
for the direct production of new qualities in the organism,
miracle fashion, by an alien environment. In a sense,
use-inheritance is a more teleological theory than natural
selection ; yet it may be subordinated to the most extremely mechanical philosophy, if in " use " environment
is held to be active and the organism itself passive.

Regarding Darwinism and chance then we have
decided as follows :—First, Darwinism asserts chance
(coexistence) in the same way in which [finite] science
ordinarily asserts it, by a mechanical view of the universe ; secondly, Darwinism has also assumed the possibility of random [non-purposeful] variations ; and on
analysis this seems to point back once more to the
same scientific assumption of distinct co-operating
forces. So far then as Darwinism really or necessarily
implies chance, it is not discredited as a science among
sciences. All of them do something similar. There are,
of course, farther questions as to the ultimate validity
of the scientific analysis, but these questions belong to
the domain of philosophy. Thirdly, however, Darwin's
phrase, " natural selection," lays greater stress upon the
element of chance than his own facts warrant. He
speaks as if the eliminating agency of a disconnected
environment were the one thing valuable. In a sense
he may be said to have made it probable that an element
of chance (coexistence) enters into the evolutionary process. But that gives him no right to say that evolution
is " due to " chance coexistence. A spark, along with
fitting proportions of oxygen and hydrogen, produces
water ; but you would throw little light upon the nature
of water by isolating one of the factors in its production,
and by describing the liquid as " due to " a spark. Salt

improves soup, but it would be a fool's enterprise to set about making soup from salt.

IV

Before we go on to test the applicability of natural selection to human affairs we may do well to ask whether, in the interpretation of physical nature, " natural selection " is not invoked in different senses. We are haunted by ambiguity. " Darwinism " is an ambiguous expression. The central contribution of Darwin to evolutionary theory was the doctrine of natural selection; yet that by itself is hyper-Darwinism; in the master's hands Darwinism means natural selection *plus* use inheritance *plus* sexual selection; these three, at any rate. So, when natural selection is used as a synonym for Darwinism, it must prove most ambiguous. May we take for granted that variation is non-telic and yet constitutes new species? Let us call this Natural Selection A. Are we to regard natural selection merely as a force that prevents relapse by weeding out possible evil specimens? Let us call this Natural Selection B. Or are we to regard it as a positive source of progress when in alliance with other evolutionary forces (telic variation, use-inheritance, sexual selection, a more general working of intelligence; all these are candidates for the position) —secondary to them, and accelerating their operation?[1] Let us call this Natural Selection C. Or, recurring to our first point (letter A), are we to leave the question

[1] The intelligent reader will easily perceive that the analysis in the text is far from being final. Is A everything? That is hyper-Darwinism. Is A something but not everything? That view might be held. Is A a logical possibility in some departments—rather unlikely to be a fact in any? That is the view argued in these pages,—and so forth. I trust, however, that all the distinctions have been taken which are necessary for our argument—in addressing intelligent readers.

open what the tendency of individual variations may be ? In that case the meaning of "natural selection" will hover between A and C. This last ambiguity is perhaps the worst of all. It leads to the insinuating or implying of A by evolutionists when they are not prepared to affirm it definitely and still less to prove it. Too often when C, or even the truism B, is established, we are asked to admit that "natural selection" has been proved. Indeed, the whole process (C) is habitually treated as if natural selection not merely entered into it but were necessarily and everywhere the dominant factor in it—as if C *were* A ; as if progressive evolution, in which natural selection plays some part, might safely be called "progress *by* natural selection."[1] It is natural selection A—the natural selection which, according to hyper-Darwinism, stands alone—that incurs the gravest suspicion of relying upon chance in lieu of reason. And it is mainly, though not wholly, natural selection A that we shall have to keep in view after this. It is natural selection A that we cannot tolerate in human affairs—least of all in morality and religion.

Natural Selection B *is a fact.*—Natural selection— A, B, or C—means primarily "struggle" and partial survival—viz. survival of the *best* (in one respect or in another). I cannot think that, since Malthus and Darwin, any one has the right to deny the existence of a selective process in nature ; and one of its effects must also be admitted—its effect in keeping each separate species up to the highest point of efficiency (natural

---

[1] The reader will please note that we are not repeating our objection, developed in Part III. of this chapter. Even although we conceded Darwin's right to speak of natural selection A, if it exists, as leading to "evolution by natural selection," we must still complain of his (and his friends') question-begging and misleading usage in speaking so not only of A but of C.

selection B). In one sense therefore, even if hardly more than a truism, we make bold (as our first step) to affirm that natural selection exists.

Natural Selection, C or A, is also to be regarded as a reality.—Perhaps the following consideration may enable us to take another step forward. Science now seems to teach that organic evolution is a fact,—that, in spite of their apparent fixity and distinctness, species have somehow grown out of each other, and, presumably, are growing still. Then, if that be so, and if a selective process among organisms is simultaneously taking place, the two processes must have affected each other (C) if they were not really one process (A). In other words; if from any cause whatever, variations capable of building up a new species are coming into existence, and if it is impossible that all organisms should live out their full span, then the new varieties will be weeded, and weeded selectively, like the rest, and this process must at least contribute something towards maturing the slowly evolving types (C; but A is possible), as well as towards maintaining in efficiency organisms of the types already constituted (B). Now, if this consideration be admitted, we may narrow the problem. We need no longer ask, Does natural selection exist? Or even, does it exist as a cause of progress? We ask, Is it the only cause? In an evolving world B implies C as a minimum, and suggests A as a possibility. Does A anywhere actually exist? Does natural selection anywhere operate by itself alone? That is our narrowed problem. That is our burning question. One school will say, Natural selection is so strong a force that we need postulate no other besides it. Another school will reply, Natural selection is perfectly credible as an auxiliary or accelerating force, but perfectly incredible as the only force. Soup (once more) is the better of a handful of salt, but you will never make salt into soup. If selection gets hold of a good thing it knows how to keep hold of it, or even how to push it on ; but it can originate nothing. It will also be possible to hold an opinion midway between these extremes. " Natural selection " by itself may be a *conceivable* cause of distinct species, yet it may be thought that other causes exist in nature which do the work more rapidly (Natural Selection A possible ; plurality of causes comes in, and Natural Selection C is the actual process).

Analysis of Natural Selection C.—The example of one concrete force assumed to be working in combination with natural selection may make our meaning clearer. Let us take use-inheritance. Lamarckism and Darwinism can be held separately, or they may be united ; but [we have argued that] since Darwin has pointed to natural selection no one can reasonably ignore it or utterly deny it. If we are to be Lamarckians at all, we must now be Darwinian Lamarckians. We may differ from Darwin as to the relative value of the two forces. Probably *any* direct evolutionary force which exists and operates must count for much more in the result than the indirect force of natural selection which co-operates with it. Nevertheless, natural selection must be producing some effect, if any process for the evolution of species is going on.

If use-inheritance is working for evolution, natural selection will back it up in two ways, distinguishable from each other if not objectively distinct. Cases of relapse by "Atavism," below the standard already reached, will be wiped out ; natural selection will be a safeguard or rear-guard to the process of evolution (Natural Selection B). And secondly, in proportion as

the competition is keener, natural selection will do more
and more to accelerate the process in a positive sense.
As between the fuller and the less full instances of use-
inheritance — the greater and less reproduction of
serviceable "acquired qualities"—natural selection will
(*cæteris paribus*) steadily award the prize to those
specimens which most fully represent the working of
use-inheritance (Natural Selection C in the proper
sense).

Another question might be asked here.   Can we
have Natural Selection A and Natural Selection C as
distinct co-operating agencies ?   Can natural selection
in Darwin's favourite sense work as a *part* of the
evolving forces in *addition* to its effect in the way of
accelerating some other force ?   Surely this can only
be the case if in *part* of the field it is the *only* force ;
i.e. if certain qualities are exempt from the operation
of the more powerful co-operating evolutionary force.
Even if you can imagine non-serviceable variations
being presented along with others, the fruit of a distinct evolutionary force, which are serviceable from the
very outset, it is almost incredible that Natural Selection
A should winnow the non-serviceable variations so as
to secure an advantageous remainder of any appreciable size.   Any other evolutionary force which co-
operates with natural selection must eclipse it as a
rival, though it may welcome it as an ally.   We
cannot *add* the working of natural selection to
the working (in the same field) of any directly
telic force.   But natural selection may *multiply* the
results of the other force — if competition is keen
enough.

Let us try to go one step further still.   As long
as struggle lasts — natural struggle — struggle *plus*

elimination,[1] natural selection is still at work. A force
may come to the birth in the process of evolution—
shall we say, of evolution by natural selection?—which
eclipses natural selection itself in importance. According to Professor Lloyd Morgan, animal intelligence is a
force of this kind. It is "far more rapid" than natural
selection.[2] Biologically, it must be regarded as an
intensifying of one valuable quality, "plasticity," or
adaptiveness and modifiability in the individual organism. The more intelligent, the more adaptable;
hence man, who possesses reason, is the most adaptable
of all animals, and has spread over the whole world.
Intelligent modifiability is inherited, as it were, in blank.
Use-inheritance [not in blank] is improbable; it seems
unlikely, says Professor Lloyd Morgan, that "habit" is
inherited in later generations as an organic "instinct."[3]
Abstract modifiability is transmitted, in the form of
intelligence; individual adjustment, helped by teaching—by the slender fund of animal "tradition"—does
the rest. Yet even here, where a new force has arisen,
natural selection is not abolished. The new force must,
I take it, *blend* with natural selection, so long as
struggle lasts. There will now be three effects of
natural selection—(1) guarding the rear—killing off
stupid members of the family; (2) pushing on the van
(killing off the *less clever* too); (3) *giving a preference*

---

[1] Mr. Sutherland may be said to plead for elimination in the human
race but not for struggle; Mr. Kidd for struggle but not for elimination.
And each of them calls his mutilated remainder natural selection!

[2] Evidently Natural Selection A is assumed—non-telic variation.

[3] I am sorry that I have failed to understand Professor Morgan's
subtler suggested substitute for use-inheritance. I cannot see how it
differs from simple natural selection (*Habit and Instinct*, chap. xiv.) Are
the modifications postulated in the organism anything more than changes
coincident with the variations in the germ? How are they conditions of
variation? Does not the selecting environment *do* everything—upon
this hypothesis of Professor Morgan's?

*to the intelligent stock as a stock over non-intelligent
or less intelligent competitors of an adjacent stock.*
This is a new point. It is another phase of Natural
Selection C. We make a separate heading for it because
it brings out most clearly the presence of intelligence
as a new evolutionary force, or, otherwise regarded, as
a new and advantageous quality. Some will describe
the appearance of a new force or quality as being due
to Natural Selection A; we have explained above why
we dislike speaking of new qualities as being "due to"
natural selection.

There is yet a further sense in which natural selection continues to work. We claim that even intelligent
animals are affected by natural selection (A?)—*at least*
in regard to their physique. It is not in animals but
only in man that we are told of an "arrest of the body."
The old sort of struggle continues in the higher brutes,
and the old lines of progress are prolonged. In one
respect therefore the old and the new forces, the slow
and the swift are added to each other; in another
respect—if we look to the growth of intelligence alone
—the two forces must be said to blend. And the
blending is in part an interference or a conflict. To a
certain extent, intelligence is so thoroughly novel as to
hamper its older comrade. If birds build nests not by
instinct but by teaching, stupid birds which would otherwise have died off will learn the essentials of life (like
stupid men), and survive ! So far then, natural selection
is thwarted. But only so far. It is not until Intelligence has become Reason that it proves strong enough
to suspend natural selection. Among the animals,
struggle still lasts; and the stupid bird will die out or
"tend" to die out in times of difficulty; though it will
not vanish so promptly as it would have done if there

had been no intelligence in the case, and if natural selection, or what is called natural selection, had been lord of all. The intelligent race will gain additional marks as against all non-intelligent races; within the intelligent race itself, the prize will still go to the best— to the cleverest *or* swiftest *or* strongest.

Can we finally decide whether or not we ought to believe in A as an actual process? Is there any region in which " natural selection " acts alone? Or—more broadly—is it legitimate to regard Natural Selection A as the great evolutionary force in nature?

It looks like a question of figures. Are there candidates enough? Is the " pluck " sufficiently severe? You may get enough of your chosen sort out of any random bunch of samples—if it is big enough. That is one view. Others again might affirm that the question is not one of numbers but of time. In (almost) endless time, any bunch that is regularly furnished will grow big enough by accumulation.

Here Mr. Sutherland gives us one shred of evidence; and perhaps we may be able to make use of it even if we do not dogmatically decide to regard natural selection as " a question of figures." The evidence is this, that the higher races in nature, when they produce offspring, follow a method of quality, not quantity. That implies that, in the higher races, natural selection, even if not suspended, has at least incomparably less room to work in. Yet evolutionary advance has certainly not been slower in these, the characteristically highest forms! This fact does not seem very favourable to what is claimed for Natural Selection A, that we ought to regard it as a reality, and perhaps as the dominant reality in evolution. For either—

(1) Though natural selection was predominant

lower down, some new mysterious force has now been disengaged, which [more than?] replaces it.   A has become C; or else

(2) If, where the best evolutionary results are gained, natural selection cannot do much, we may hesitate to believe that it produces much effect at any part of the process.   There must be other forces; telic variation and use-inheritance are candidates.   Not A anywhere; C everywhere.[1]

To repeat our conclusions then; natural selection (A) is certainly not the only principle of evolution in nature. It is very doubtful whether there is any part of the field where it stands alone (whether Natural Selection A exists), though it seems metaphysically possible; *i.e.* the supposition seems to be sense and not nonsense. On the other hand, it is certain that the law of natural selection (B and especially C) is at work, with large effects, in every part of what is strictly called Nature.

V

In the next place, we have to approach the central part of our subject, by asking how far natural selection is applicable to human evolution?   Here as elsewhere the burning question is whether Natural Selection A can be applied to human affairs.   But we must keep all three forms in view—A, B, and C.

---

[1] An odd suggestion offers itself. Can we combine Mr. Sutherland and Professor Lloyd Morgan? Can we hold that the higher animals are able to advance with less help from natural selection, because they have more help from their intelligence? One must note a distinction; physical evolution *by means of* intelligence is not identical with *the evolution of RATIONAL intelligence*, which Drummond, etc., believe "arrests" the body. Higher brute races are certainly intelligent, and (I suppose) are certainly evolving physically.—Dr. Mellone (*Studies*, etc.) puts a different construction on the whole question. He inclines to assume a psychical factor in all evolution, even of plants,—on Wundt's view, that plants are descended from animated ancestors! This is very un-Darwinian.

We must also distinguish between the biological view of man—where natural selection is most likely to be at home—the sociological view, and the moral view. Man is still an animal, an organism, though he is also a citizen and a moral agent.

First then, biologically, does natural selection apply to man ?[1]

The Struggle with the Beasts.—When we read in the Bible of man's dominion over the creatures, we naturally think of domesticated animals, or of those wild species which man—and woman—make use of for food, clothing, ornament, etc. But man's supremacy over savage and powerful animals is a far more wonderful fact. There must have been a period of sharp conflict. Even in the Old Testament (to quote it again) we have traces of the dread lest wild beasts should gain the upper hand, and make human life heavy with torturing anxiety. The conflict ended however in a decided victory for the seemingly weak race of man. His dominion became a reality. His fear and the dread of him affected even the most formidable among his animal subjects. It was fixed that his life should follow its regular course, unhindered on any great scale by the evil beasts. They could only carry on a guerilla warfare. When they slay a man, it is an " accident," and, in spite of such exceptions, the human race marches bravely onward. Men have emerged from this struggle for existence. The struggle continues

[1] The " Arrest of the Body " seems to imply that physical evolution is at an end, and therefore that the force of natural selection, which makes for evolution, is also at an end. And in the closing chapter we shall quote names of high authority who deny that natural selection applies to man or at least to civilised men—Darwin, Professor K. Pearson, Professor Lloyd Morgan. But it may be well that we should here look for ourselves into the details, and form our own judgment.

beneath them; but they, with whatever limitations
and exceptions, are victors, and champions of the
world. Concurrently, they have learned—again, with
certain limitations and exceptions—not to struggle
à outrance against each other. There follows from
these two attainments (once again, with some strange
and saddening exceptions among the lower human
races) that man has the awful prerogative and solemn
privilege of dying a natural death. Such a thing is
rare in the animal world; but men drink their cup
of pain to the last drop, and pass, it may be, with
unbandaged eyes behind the veil, into the unseen.

Famine.—Emergence from struggle with animal
competitors *may* signify nothing better than a liberty
to die of famine. Natural selection does not govern
the physiological development of men, for they have
not overfilled the world; but a local and temporary
over-population not infrequently arises, and famine
follows close upon it. Civilisation ought to have other
means of coping with such an overplus; nature treats
it as a normal case of animal superabundance, and
falls to selecting again by the old eliminating methods.
The human harvest is weeded; the strongest survive,
weakened—probably not permanently injured; others
succumb. Here then is natural selection at work
among men, and conceivably Natural Selection A, if
Natural Selection A anywhere exists. Of course it
will be much hampered, more hampered than among
any of the animals, by the comparatively low rate of
fecundity in man, though famine goes a certain way
towards remedying that. Among the higher animals,
as we saw, evolution has continued no less markedly
than with the lowest, and we decided that some other
factor making for progress must be in operation there

besides natural selection (A). We have no similar assurance that biological evolution in the sense of progress is continuing among men — have we not heard of "the arrest of the body"? If evolution continues it must owe its strength to something besides the recurrence of famine. That is not frequent enough. It does not "eliminate" severely enough to enforce progress, even if it tends that way.

On the other hand, famine has been no rare thing among savages—no rare thing even in the history of the civilised world. For good or for evil, elimination has acted on mankind through this agency; and yet every civilised government, even the hardest, is ashamed of famine, and overwhelmed with a sense of defeat when its people are starved. Probably, if famine were allowed to stalk the world unchecked, we should see the selection of a corresponding physical type in the human races ; a low type ; prolific ; tenacious of bare existence ; never rising much above the margin of subsistence and possible survival. The upward path lies elsewhere.

Pestilence is another eliminating agency which takes the weak and spares the strong, though it is much more likely than famine to leave behind it dangerous and enfeebling "dregs" in those who recover. It has been supposed, indeed, that the Jewish race owes some of its health to the fact that the hideously insanitary conditions of the mediæval ghettos killed off the weak. Strange if the most sanitary and the least sanitary conditions should alike result in producing a healthy human type ! But there seems every probability that the Jews were already one of the toughest of human stocks when they entered that furnace. They emerged hardened still further ; ordinary human races might have succumbed. If *we* fell back on "natural selection," not sanitation, to make our people healthy, we might succumb !

It is not to be denied that pestilence is relatively advantageous. If the world must go on under conditions of filth, it is better for the race that the resulting diseases should blaze up in intermittent epidemics, carrying off the weakest, than that they should linger on as a chronic leaven of weakness and pain, tainting the whole race. But it rests upon us to find a better system than the serviceable pestilence. I say again in all seriousness, if we selfishly fell back upon *laissez faire*, natural selection might eliminate us *all*. Civilisation may well have softened our fibre in some respects ; and *homo sapiens* has no title-deeds to life guaranteeing him its continuance. Of all conceptions of the end of the (human) world, none perhaps could be more ghastly than the vision of a race dwindling away, from vice, from self-indulgence, from inherited disease—a race that could not rise to the responsibilities of reason and conscience, but called "sauve qui peut" when danger came, with the result that from the ensuing stampede none escaped without fatal injuries. If we fall too low, wise nature will simply stamp out all of us.

And yet we have in pestilence, while it lasts, an accessory selecting agency (Natural Selection C), with the drawback noted, that the monster leaves the mark of his talons upon many who escape with their lives.

Vice.—Mr. Sutherland lays much stress upon the excellent results due to elimination of the vicious. This is of course Natural Selection B, and nothing more. Prolong to infinity the elimination of vicious persons— will that develop virtue ? At least it would not, upon any view, improve its quality. Another favourite idea is that any special vice, if left unchecked—*e.g.* drunkenness—will burn itself out by natural selection. Dr. G. A. Reid's " Present Evolution of Man "[1] argues for this pleasing possibility. Surely this is folly. Men are not of distinct kinds, as the old Gnostics supposed. We can *acquire* qualities by developing their germs ; we can make the transition from the class of the sober to that of the drunken. It is only too easy ! Frightful as are the penalties of such vice, when have they proved sufficient to counteract the charms of jollity and good fellowship, and of a " moderation " which so easily becomes immoderate ? Mr. Sutherland himself implies that each generation or two develops its own criminal class, its own profligates. Assuredly upon that point he is credible. Human nature is versatile, and man is weak ; a new crop of drunkards may easily be grown as the old ones die out. If you leave everything to natural selection, that is how the world will go.

Crime, or human justice punishing crime, is also a form of Natural Selection B. Eighty or a hundred years ago criminals were " eliminated " wholesale, with little profit to society ! The problem of human advance proves unexpectedly complex. Brutal violence on the part of the law provoked more crime than it repressed. Even at the present day, however, we do some " eliminating." We hang a few criminals, and we seclude others, both men and women, for long terms of imprisonment, during which terms at least it is impossible for them to produce offspring. We may attribute these results to *Natural* Selection if only in this sense, that the reduction or checking of population was not the design of our criminal law, but an incidental consequence.

It is a favourite idea with some students of society

[1] Quoted in *Habit and Instinct* and elsewhere.

that "the sterilisation of the unfit" ought to be carried very much farther. Theoretically, one is tempted to sympathise with the opinion, but it is doubtful whether any such mechanical methods will do much for human welfare.

War is among the strangest and saddest of man's institutions. Systematised violence and wholesale slaughter are new things in the animal world. War has been immensely widespread and potent in the course of human development. Socially, we saw that, as between community and community, war has often done good. In early days the best fighters are generally the best tribe; and war has not infrequently become a pioneer of civilisation. But, alas! at what a cost! Morally and socially, the cost is beyond reckoning. And biologically, or in its bearing on individuals, war has usually snatched away the fittest and left the weak or the cowards to become the parents of the next generation. During early ages, while individual valour counted for much, war exercised some influence in the way of selecting the best—backed as it was by a sexual selection; redcoats have always charmed the gentler half of the race. But in recent times the characteristic effect of war is downright evil, as when the Napoleonic campaigns (it is supposed) lowered the stature of the whole French nation. War is a selecting agency of great influence turned upside down.

Religious Celibacy has possibly had more consequences, good or bad, in its moral and social than in its physiological bearings, and it is a historical rather than a natural force; still it may be mentioned here for convenience. When you take account of Buddhism as well as of Christianity, you perceive that religious celibacy has been a phenomenon on a vast scale, and with a gigantic influence, like war. Like war, too, it has selected steadily in the wrong direction. The best and finest spirits were withdrawn from family life; the inferior types were left to perpetuate their qualities in offspring.

We see then that famine may possibly show the working of Natural Selection A within narrow limits; pestilence and disease, if they do anything positive, must be ranked in Natural Selection C, as mere accessories to some better force; the fatal or sterilising consequences of vice and crime do no more than protect the rear—Natural Selection B; war and religious celibacy select, but select pretty steadily on the wrong side.

It does not appear therefore that natural selection achieves much for progress, or much even for advance of any kind, in any one definite direction, within human affairs, when viewed biologically. The view of natural selection implied in the doctrine of the "arrest" of the human "body" is upon the whole confirmed.

But, if it were the best thing in the world, mankind cannot make use of natural selection. We must keep each other alive and well, as far as we may; humanity insists upon it. In point of fact, the civilised races are putting their chief reliance, for biological progress or safety, upon forces of a very different kind. There is first—for we are speaking here of man's physique— the provision, by laws and by administration, of a sanitary material environment; next comes the advance of medical skill, the diffusion of medical and sanitary knowledge, public opinion, law (requiring and forbidding certain individual acts), morality, religion. That is the line we must move on, whether we like it or not. And we have no reason whatever to suppose that we should get better results by "following nature" in a more brutal fashion.

In sociology Mr. Benjamin Kidd has claimed that all our salvation lies in natural selection, failing which "panmixia" entails retrogression. This is really biological rather than sociological doctrine, and probably or certainly it is bad biology. There is little or no true struggle for existence among human beings; thank God for that! Reason and our moral nature make it impossible; and yet we seem to have escaped retrogression. Mr. Kidd dwells on the necessity of struggle, while he says nothing about elimination; and he applies his supposed biological truths directly to the human animal. Reason is held to affect the process chiefly in a dangerous way. It makes men clever, no doubt, but it makes them too selfish to struggle in the interests of the species, unless religion had come in to keep us up to the mark. Social evolution therefore depends on (what is called) natural selection, *minus* reason, but *plus* religion. So Mr. Kidd tells us.

We should say that it depends upon reason *plus* morality *plus* religion. And since it depends on reason, it depends on those who have reason most fully, and yet are brothers to those who have least of it; in other words, social progress depends upon great men. We lesser men stand on their shoulders; as reasonable beings, we share in their discoveries. On their side they are not independent of us, the little men. Even the ruthless Napoleon Bonaparte is said to have made that confession. "Why," he asked of David, "have you put those tiny troops and guns into the corner of my portrait? I ought to be alone." The embarrassed painter apologised as best he could. He thought that

the sketch of the army in the background had a historical interest, etc., etc., when Napoleon, having recovered
his good temper, remarked, " After all I owe a good deal
to these worthy little men." Well might he say so!

Finally, among the different human provinces, we
have the assertion that natural selection prevails in
morals.

Prof. Alexander alleges this of moral *ideas*. They
struggle against each other, and the fittest survive.
Stripped of metaphor, the meaning is that free discussion is a condition of progress in moral thought. Surely,
it is *one* condition. But it is a psychical condition; it
implies reason; it implies the power of the great man
to indoctrinate others.

Mr. Alexander has not affirmed literal natural
selection. It was impossible that he should, though in
some respects he has gone too near it, and has thus
exaggerated the mutual repulsiveness and exclusiveness
of distinct types of ethical thought.

But, after all, is not the main point here just the
one which Mr. Alexander (following Darwin) takes for
granted? Whence come the new varieties? In dealing
with morals, at any rate, this is all-important; and in
dealing with morals, at any rate, this cannot be answered.
Even the victorious analysis of the evolutionist is baffled
here at the central point—

> A spirit breatheth, and is still;
> In mystery our soul abides.

What one can say about new and sound moral ideas
smacks painfully of platitude. Sometimes they may
be championed at first by moral eccentrics. But usually
the teacher will be well rooted in the past, drawing

from it his best strength, seeking not to destroy, but to
fulfil.   Yet even he is likely to be proscribed, insulted,
hated, and perhaps killed.   Not till after his death will
men recognise the truth of his words; then they will
quote them against his successors.

Mr. Sutherland deals not so much with the growth
of moral ideas as with the history of moral behaviour
and the growth of character.   The doing of what is
right concerns him rather than the knowledge of it;
these are distinct problems.   His belief is that we
grow better because the vicious and sensual and
violent die off, leaving few children.   If there is any
other evolutionary factor, it is so paltry in extent that
we may safely disregard it.   Natural Selection B is
to smuggle in Natural Selection A concealed in his
pocket, or is to disguise himself as his big brother.
There is no such thing as a new life for the repentant
sinner, and there is no influence (to speak of) between
man and man.   The filthy remains filthy still, and the
righteous remains righteous still.   We are born good,
or else we are damned into the world.   Elimination is
first among moral forces; the rest are nowhere.

What is a truer theory of man's advance in actual
goodness ?   We help each other—by influence, example,
magnetism.   And inwardly we are drawn or driven to
righteousness partly by the bitterness of sin, partly by
(not the pleasures of virtue, but) the beauty of holiness.
It would be impossible to say which has the more power.
The great inspiring personality who helps the multitude
of little lives may be unoriginal and hackneyed in
thought.   It is the glow of spiritual goodness, *plus* a
mysterious personal endowment, perhaps of the nature
of sympathy, that constitutes greatness and efficiency
in this department.   But the " worthy little men " are

quite as important here as the leaders. Mr. T. H. Green has told us that the Napoleonic wars were able to do some good, as well as mischief, in the world, just because of the courage and loyalty of the millions of private soldiers who were the victims of one man's ambition. Faithfulness is the greatest of the virtues. Nor must we forget the stored wealth of the past in the form of moral institutions and traditions.

We have one proof of the all-sidedness of Jesus Christ in this, that He is both the supremely original moral teacher and the supreme personal influence. He so crossed the currents of dignity and respectability in His age that dignity and respectability, feeling "what such men call the 'necessity' of putting him to death," tried—strange endeavour!—to "eliminate" Him! Yet without strain or manifest extravagance the view can be advanced that it was His glory to put the great moral commonplaces into circulation as "current coin." We go to Him for "sweetness and light." He is the truth. We go to Him for transforming warmth, and He makes our cold ideals live, and melts our hearts.

VI

Natural selection then does not rule within the sphere of reason. We may now face the question, whether it can be said to account for the first emergence of reason and morality?

One is reluctant to admit this. Yet it seems as if there was almost the same warrant for ascribing the emergence of reason to natural selection as for imputing to its agency any other new thing that arises in the course of evolution. Darwin's language we have pronounced ill-balanced. Natural *selection* does not *create*.

In speaking as if it did, Darwin ignores a co-operating factor of even greater consequence, the capacity for aggregate variation in the material. Moreover, selection out of non-telic elements seems possible, if at all, only in the lower ranges of evolution, where fecundity is greatest. Yet it may be held that reason emerges *by means of the process called* natural selection, and by means of a process in which natural selection, *i.e.* struggle and elimination, have certainly played some considerable part. On the assumption of evolution all along the line, it is implied that [what is on the surface] a natural process has led up to the spiritual forces of morality and reason. Being a natural process, it has never wholly shaken off the influence of elimination.

Of course, if Mr. Wallace is right, that there was a special supernatural intervention when reason appeared upon earth, it will not do to say even in the most guarded sense that natural selection created reason. But this *quasi* "miracle" is doubtful. Mr. Wallace himself has laid the greatest stress upon the preservation of reason by natural selection. We prefer his teaching on that point, with all its difficulties.—Why (for example) have the irrational races not died out? Can we hold that the race nearest man, yet irrational, died out in competition with him? Perhaps that is why it is so hard to find traces of the missing link. Presumably competition is always keenest between adjacent forms. Consequently, defeated species may disappear outright, and their disappearance may explain that semblance of a gap between the nearest existing species which is so noticeable in many parts of nature. I do not know whether this suggestion has been made before. If not, it may be offered for what it is worth to those who are interested in defending natural selection.

VII

The view now sketched of natural selection—that it is a real force, but strictly limited—has been outlined in a spirit of sympathy with idealistic philosophy. Yet it is opposed to the views of several distinguished Hegelian idealists. Some of them would say that it goes much too far in commendation; others, not far enough.

Professor Ritchie endorses natural selection without putting any limit to its application. It seems to give him all that he needs. There is evolution in Darwin, and there is evolution in Hegel; therefore natural selection accounts for everything, or at least it does so *mutatis mutandis*. We have tried to show in detail what the mutation is, and it is pretty extensive.

(On the other hand, Professor Ritchie, as social philosopher, takes the opposite view, holding that reason has transformed the whole evolutionary process which it has touched.)

Dr. Stirling and Mr. Sandeman, if I understand them rightly, regard natural selection as a piece of showy but flimsy thinking, that crumbles away as you handle it. They would deny that it explains anything, or that it applies to any part of the cosmos.

Mr. Sandeman [1] believes thoroughly in the teleological character of organisms, and finds every existing species too perfect and harmonious and balanced to think of "bettering itself." Instead of the realistic vision of cosmic horrors, he has a poet's vision of peace. He is not content with excluding absolute unfitness, but insists on denying even relative unfitness. "Whatever is, is right." It exists, it has survived; it

[1] *Problems of Biology.*

triumphs! Like the apostle Paul, Mr. Sandeman bestows more abundant comeliness upon our uncomely parts. With great force and penetration he observes [1] that inherited rudiments have not been inherited as ready-made rudiments; they have been built up along with the rest of the organism, taking their full share in the reciprocities of organic growth. *Ex hypothesi*, what is really (and not merely apparently and externally) useless must long ago have disappeared under the fierce strain of struggle for existence. Yes, very good. Nothing is absolutely unfit. The most rudimentary part must discharge some obscure physiological function in the rhythm of life. But are we really to suppose that the human body would be wrecked and ruined if (say) the *ilium cæcum* were somehow and safely evolved out of existence—as the surgeon on emergency may cut it out? If such a body arose, would it not be a better body than ours, so far as hitherto evolved? Is it unthinkable that nature should improve in this fashion? Is not the whole living world relatively fit, indeed, but also, in many important details, relatively unfit, and is not an aborted organ very plainly marked by nature as, in one most important sense, unfit?

Mr. Sandeman presumably implies the absolute systematic perfection of the whole universe as well as of each individual organism, and presumably affirms this postulate on metaphysical grounds. Even without repudiating it, we may urge that the idea is not applicable off-hand to the world of nature. Men will not readily surrender that dynamic view of nature, as a great and incomplete process, which Darwin and other evolutionary thinkers have taught us. The optimism

[1] This is a valuable corrective or supplement to Professor Ritchie's criticism of Dr. Reich, *Darwinism and Politics*, pp. 124, 133.

of Mr. Sandeman's own creed does not force us to affirm the perfection of the individual organism save as a part in the process by which the perfect whole evolves itself.

Dr. Stirling, on the other hand, finds the individual too poor for the work required of it in Darwinism. So far as I understand his position, it has two elements. It nails Darwinism to the assertion that variation is casual (as it were, causeless). And while repudiating such "casual" difference as a source of progress or as a possible beginning of specific types, it alleges the existence of the casual element under the name of "individual difference," which seems to be in Dr. Stirling's[1] view all but aimless and all but causeless.

Perhaps the meaning is this. Every individual differs from every other member of the species. The difference does not affect the specific type or pattern; it neither augments nor lessens efficiency. Each is a man, a fish, a frog yet each has its own peculiarity, its, so to say, *casual* peculiarity, indifferent to the specific type. To get species—law—rational system out of this most casual, most non-systematic of all things in the cosmos—that is the alchemy of Darwinism; out of a brew of chance, to distil pure reason ! The casual difference *is* just the drop of unreason, of brute matter, dropped into the specific type in order to make it down into a new individual. This, so far as I can conjecture, is Dr. Stirling's meaning. No summation of *individual* peculiarities can ever amount

[1] I am thinking of *As Regards Protoplasm* and *Darwinianism*, but mainly of Dr. Stirling's *Gifford Lectures*. The very acute mind of Dr. Stirling suggests innumerable objections to Darwinism. We have only dealt with what seems to be the central point—the denial that the alleged process is reasonably thinkable.

to a *specific* difference.   The things are heterogeneous in their very essence.

Now I will not attempt to criticise the metaphysics of this.   But I venture to assume that such thinking lies too deep for science.   No biologist would hesitate to speak of "identical twins," or would admit that heredity acts differently at each birth, merely in order to put itself metaphysically in the right in its act of bringing into the world a new individual.   If "heredity" should not differentiate individuals, "environment" would speedily do so.

On the other hand, I submit that the "casual" variation which science speaks of is found, when science sifts its thoughts, to be one which—whether actual or only possible — might quite well conceivably, by cumulation, amount in time to a new specific type.   Of course there are difficulties in detail under Darwinism. But is Darwinism ;metaphysically incompetent?   Does Natural Selection A outrage common sense when you understand its terms?   I think not.   It is certainly limited in range ; it possibly exists nowhere in nature as an actual process ; Darwin's name for his theory may be misleading ; but surely the theory is conceivable.

Finally, let us observe that, even as a fiction, natural selection might be serviceable, though the truth were merely that species *are things which might have resulted from infinitesimal changes in endless time.*   Even on that view "natural selection" might be a fruitful guide to investigation, not a blind alley. *Per contra* the fruitfulness of natural selection as a theory does not in itself certify it to be a true theory, whether in whole or in part.

PART IV

HYPER-DARWINISM—WEISMANN, KIDD

CHAPTER XVIII

A "FAIRY TALE OF SCIENCE"?

An intenser assertion of struggle—Not on ground of experiment ; evidence is ambiguous—On ground of a theory of heredity—Darwin's theory (Pangenesis) assumed derivation of embryonic qualities from qualities and tissues of parental organisms—Use-inheritance possible or probable on this view—But "Atavism" forced the concession, some "gemmules" had passed on undeveloped from earlier generations till they found their chance —Galton's figures for resemblance to ancestors—Hence theories asserting "continuity of the germ plasm"—Parable of the Hierarchy—Galton ("Stirp") does not absolutely deny the possibility of use-inheritance— But in Weismann's earlier and more consistent views, founded on by Mr. Kidd, amphimixis is the only cause of variation—Extrusion of one of the "polar bodies" securing (?) non-identity of all offspring of same pair— Permutations and combinations of qualities of unicellular organisms— Nature selecting fittest adults, and in them best germ plasm—Parable of the suckers—Of the Nile—No new quality arises, but amount of each telling quality increases—Qualities arose originally, Lamarck fashion, from environment, when unicellular life lay open to its pressure—Unicellular organisms (propagating by fission) and germ plasm are potentially immortal —Correlation alleged between sex and (natural) death ; now sex is absent from the unicellular world—Natural selection might account for the *predominance* (if not *origin*) of sex if Weismann would assume the necessary competition—Romanes alleges that natural selection might account for predominance of habit of dying natural death ; but would not *death by violence* sufficiently prevent any race (immersed in the struggle) from falling into wholesale decrepitude ?—*Origin* of sex and death a mystery ; or "chance" variation ! or effect of molecular constitution of germ plasm !—Weismann's appeal to "natural selection," while he denies "struggle," is metaphysical in the worst sense—Recapitulation, and note of some of Weismann's changes of opinion before 1893—Especially this change : ENVIRONMENT MAY DO SOMETHING TO MODIFY GERM PLASM !—Making true use-inheritance conceivable, though not inevitable—Mr. Kidd is anachronous—Panmixia, the absence of natural selection, is held to involve indefinite retrogression ; important ; questionable.

ALTHOUGH we have passed under review a reaction from
Darwinism, on moral grounds, or in the moral region,
yet the theory which in recent years has excited most
attention, both popular and scientific, is not a qualification of the Darwinian doctrine of struggle, but an
intensified assertion of it. Weismann, like the young
Rehoboam, meets all discontent with a stiffer front and
a severer policy. "My father chastised you with
whips, but I will chastise you with scorpions." Darwin
laid a terrible emphasis upon struggle for existence;
but he admitted other causes of progress, such as sexual
selection and use-inheritance; Weismann admits no
cause of progress whatsoever, except struggle for
existence; no selection of the beautiful by the instinct
of sex, and above all, no inheritance of acquired
qualities. Such is Weismann's position; a scientific
position in regard to technical questions of biology,
held by a competent and highly distinguished, though
also a highly speculative man of science. But the
position manifestly involves or suggests inferences regarding human progress: and these are worked out
with devout fidelity, and with much ability and
knowledge, by Mr. Benjamin Kidd.

Primarily, the question between Darwin and Weismann is one of fact. Does experience confirm or does
it refute belief in the inheritance of acquired qualities?
Unfortunately, this question like many others is more
easily put than answered. Romanes tells us (in the
preface to his *Weismannism*) that he himself, acting
under Darwin's immediate direction, instituted a long
series of experiments on the point; but that the results
of these labours, which extended over several years,
were never published, because the experiments "all
failed," *i.e.* presumably, they yielded incurably ambi-

guous results. "Nothing is so deceptive as facts"; the
same facts are capable of such different interpretations.
Apparently, Weismann has shown that the range of
the "Lamarckian factor" was grossly exaggerated.
To that extent facts openly support him. Whether he
has proved that use-inheritance does not occur at all
is another question. The non-inheritance of mutila-
tions, even such as have been persisted in by custom
through age after age—Chinese foot-binding is a notable
instance—furnishes a strong argument in Weismann's
favour. And even hostile evidence can be robbed of
much of its strength. Are there not blind fish in
the mammoth caves of Kentucky, and in similar
caverns elsewhere? Have not preachers freely used
this illustration of the bad results of evil *habit?*
Yes; but if there was no premium on eyesight, fish
which "happened" to be born blind would have an
equal chance of living and begetting progeny with
fish that saw. Give it time, and natural selection—
or in the opposite case, panmixia; the cessation of
natural selection—will produce all the results com-
monly attributed to use-inheritance. Use-inheritance
would be a much quicker process; but have patience
with natural selection (or with panmixia), in a few tens
of thousands of years it will do all that you require.
Other suggestions are that, in dark caves, the fish
which put part of its physiological capital into a superfluous sense would be positively disadvantaged by its
eyes in the struggle for existence. Having wasted its
resources on an inherited habit of luxury, it would
fail in securing the necessaries of life. And again,
Professor Ray Lankester has suggested that the fishes
with good eyesight would find cracks by which they
could swim away, leaving behind them only the blind

or purblind. If any of these were suffering from mere
accidents to their eyes, they would of course on
Weismann's hypothesis beget a progeny having eyesight. But, if any had their vision congenitally dim
or dark, they would become the parents of those blind
fish which we know.

Thus the facts give an uncertain answer, and we are
driven to make a statement of the blendings of fact with
hypothesis which have been championed on one side or
the other. Theories of heredity are invented to suit
the facts, so far as known, but they lie far beneath the
strata where verification is possible, at least in the
present state of our knowledge.

The simplest and most natural assumption is that
the embryo, or its antecedents, spermatozoon and ovum,
owe their qualities directly to the parental organisms.
"The owl comes from the egg, but likewise the egg
comes from the owl." And this natural assumption
leaves the door open for the farther assumption that
acquired qualities will be inherited. I do not see that
it compels us to hold that view. An acquired quality
may be (as it were) only skin deep—having no reaction
on the inner life of the organism—not stamping its
mark there, and therefore not stamping its mark on the
offspring, which reproduces that inner life in a new
generation. If living shells, transported from a northern
sea to the Mediterranean, assume the same bright mark-
ings found in native Mediterranean forms, who will
believe that the change, however conspicuous, is the
same thing as transition to a different species? They
are still essentially the same, and their offspring will be
essentially the same, bright if developed in the Mediterranean, dull if developed in the north. But that the
deeper qualities of the parental life are all reproduced

by it in its offspring—transferred from it to its offspring
—seems to correspond best to the proved nature of an
organism as a unity or system, in which all parts are in
reciprocal intercourse, and the whole determines all the
parts.   One mark or outcome of this reciprocity will be
the alternation already spoken of, owl from egg, egg
from owl.

Darwin represents this natural assumption; but as
it occurs in him it is attended by some peculiarities
due to modern science.   Science is bent on finding a
mechanical cause for every mechanical result, and on
eschewing mysticism.   The effort is laudable, if it can
be carried through without injustice to the facts of
organic life.   But it results in a singularly self-confident
materialism; or so one is tempted to think.   It analyses
the organism into a bundle of qualities, and postulates
a separate speck of matter or living vibration for each
quality distinguishable from the rest in human thought
and speech.[1]   The description applies, among other
hypotheses, to Darwin's "provisional hypothesis of
pangenesis."   According to Darwin's view, each part
of the adult and vigorous organism gives off extraordinarily minute "gemmules."   These work their way to
the parts of sex, and pass on as "packets," one paternal
"packet" blending with one maternal "packet" in the
embryo, and gradually reconstituting a body, each
gemmule helping to build up an organ, or limb, or
tissue, like that from which it sprang.   Facts, however,
insist on a serious qualification, the facts known as
atavism.   Often, or always in some features, the child
resembles a grand-parent or remote ancestor more than
it resembles either parent.   How is this to be explained?

---

[1] This criticism is urged very tellingly by Mr. George Sandeman in
his *Problems of Biology.*

Again we are forbidden to fall back on mysticism, or to omit the discovery of a physical and mechanical cause. There must be *gemmules* from far-away ancestors developing in each child. It follows that in each embryo some gemmules must fail to develop, but, instead of perishing, must pass on as gemmules, with all their latent qualities; must enter with other gemmules into new packets constituting ova or spermatozoa, and must find their chance of development in a later generation by a triumph of atavism. Thus it is only partially true in Darwin's opinion that the parent organism and the reproductive material are in full sympathetic reciprocity. Distinct part of the latter, according to Darwin, though *in* this generation is not *of* this generation; though living in, and by, the living body of the adult of to-day, it owes its origin to other bodies, whose qualities it hopes one day to reproduce when its chance arrives. The owl comes from the egg, but the egg comes only in part from the parent owls. Another distinct part of the living embryonic substance owes its being to older birds. Mr. Francis Galton, great experimentalist and statistician, has arrived at a formula for the higher races. One-fourth he calculates belongs to each parent, one-sixteenth to each grand-parent, and the remaining aliquot part of one-fourth, I presume, to remoter generations still.

It must not be supposed, however, that Galton agrees with Darwin in believing in pangenesis. His position is much more nearly that of Weismann. He can only hold that one-fourth part in each of the offspring is (on the average?) *like in quality* to the father or mother, not, as Darwin might do, that the child owes its being and nature in the proportion of one-fourth to the father, and the same to the mother.

By a fuller consideration of the problems of atavism, and by a growing hesitation to admit the inheritance of acquired qualities, doctrines of the continuity of the germ plasm have gained in popularity and acceptance. There are difficulties about the facts. In certain animals it appears that, at a very early stage in embryonic development, part of the segmented ovum is differentiated for reproductive purposes. Here then the parental germ may be styled continuous with the germs which are a preparing in the reproductive tissues of the growing embryo. But in most cases it is a long time before we reach specialised reproductive cells. The germ cells seem to be derived, if only at this early stage in development, from somatic cells, and continuity with the past seems to be disproved in favour of reciprocity in the present. At this point, therefore, Weismann and others take a deeper plunge into sub-microscopic minuteness and unverifiable theory. They cannot prove continuity of germ *cells*, but nothing can hinder their asserting continuity of germ *plasm* or the like, *i.e.* continuity of the invisible substance, believed to form part of the contents of [reproductive] cell nuclei, and to be the vehicle of hereditary qualities.[1] On this view of things we must alter our parable. The owl comes from the egg—true ; but the egg (the microscopic living embryonic ovum) never came from the owl—never ; the owl came from the egg, and the egg came from the egg. The living hereditary substance, the assumed carrier of the qualities of heredity, is called by Galton " Stirp." Weismann calls it " Germ plasm," subsequently " Idioplasm," and later on introduces

---

[1] The phrase (in the allied form, " continuity of the germ *proto-plasm* ") is not of Weismann's coinage, but goes back to a previous writer, Jaeger.

farther refinements and subdivisions. If we may take
an ecclesiastical analogy, the ordinary doctrine of organic
reciprocity corresponds to the Protestant doctrine of the
Church. The ministry are specialised organs of the
Church, kindred to all other parts of the living Church
tissue, capable, if the need arises, of being replaced by
any other part without serious damage to the true life
of the Church. On the other hand, "continuity of the
germ plasm" corresponds to the High Church doctrine
of apostolical succession. Age after age the Church is
made or created by the hierarchy, but the hierarchy is
never made by the Church; it is made by the ante-
cedent hierarchy. There is no reciprocity, there is no
fellow-ship, but aristocratic superiority on the one side,
and absolute dependence on the other. If the hierarchy
perishes, or is interrupted, everything is lost. A strange
belief surely! Yet who knows? If certain views are
biologically correct, the High Church school of Christians
may claim to be more scientific than any others. But
are these views proved, or even permissible?

In their full (and *quasi*-High Church) severity,
these views are to be found only in Weismann's earlier
writings, where he develops his more characteristic positions. "Stirp" always differed from "Germ plasm";
for Galton always admitted a certain modified action of
"use-inheritance" or "the Lamarckian factor." And,
along with other changes registered by Romanes in
1893, Weismann had by that time withdrawn his former
doctrine of the "absolute stability"—so Romanes puts
it—"of the germ plasm," and had come over to Galton's
view, according to which the influence of environment
in originating variations, and so contributing directly to
evolutionary progress, while slight, is yet not to be
denied. However, the earlier form of Weismann's views

must be regarded as the more coherent and original. It is almost as interesting as a fairy tale, if possibly not much truer. To an outside critic, at any rate, Weismannism seems to have grown latterly after the manner of a false hypothesis, not after the manner of truth. It has modified itself endlessly by adding on ingenious epicycles. Instead of leading to new generalisations and broad views of things, the changes have made it complex and artificial-looking. True or false, the older Weismannism is at any rate clear,—clearer than the new. And Mr. Kidd's sociology seems to appeal to the Weismann of 1893, or of still earlier years, not to the author of the later more hesitating statements.

At first, then, Weismann had held that germ plasm was never affected by the life of the organism in which it was temporarily lodged. It was perfectly continuous, absolutely stable.[1] Yet varieties occurred ; for evolution occurred ; and there was no cause of evolution except natural selection ; and natural selection could only work upon given materials. Whence then did varieties proceed ? From amphimixis and from that alone ; in other words, from the processes of bisexual parentage. There was "nowhere else" for variations to come from on this early and rigid theory of Weismann's ; and the theory threw a delightfully definite and clear light on the cloudy problem, what is the origin of variations ? No doubt there was a difficulty here. If individual variation is due simply to parentage, why are not all the offspring of the same pair facsimiles of each other ? Can science clear up this mystery ? Weismann in his early phase explained it by the extrusion of one of

---

[1] Apparently the phraseology is Romanes'. To a layman it looks tautological. Romanes himself (pp. 49, 86 of *Weismannism*) seems unable to keep the two terms distinct in their application.

the two *polar bodies* expelled from the ovum shortly
before—or more usually shortly after—fertilisation. I
do not know that I understand this.   Up till now,
germ plasm has been described as so continuous or so
stable, that it has threatened to make all the offspring
of the same pair identical with each other if the two
parental germ plasms are simply added together. But
now, wise nature casts away half the qualities or potentialities of the germ plasm, when it throws away half the
substance, and the dividing line is drawn *at random*,
or at any rate, is never twice the same. Weismann's
later view, to which Romanes had thought that he was
bound to come—and on which Romanes looks with less
disfavour—seems to involve the same difficulty. How
can cell segmentation divide the germ plasm into different potentialities, corresponding to differences exhibited
later in the different members of the litter or family, if
we are to hold to the high stability of germ plasm? Or
how on earth can we reconcile this with the doctrine
that amphimixis is the only source of variations?
Moreover, are we to understand that germ plasm,
" which grows very rapidly," never grows at all, or
never segments at all, after birth?   If it did, apparently
it would be constantly changing its qualities. It would
be highly *un*stable.[1]

Nature then, according to Weismann, had been
playing an immense game of permutations and combinations, if not since the dawn of life itself, yet ever
since the first origin of multicellular organisms, whether
plant or animal. All of these become unicellular at
the beginning of the embryonic process, when the new

---

[1] The polar bodies had to serve as the explanation of a second
difficulty—one of size.   It also is mysterious.   On it also Weismann
has changed his ground.   And by that change also he secures greater
approbation from Romanes.

life is constituted by fertilisation. And therefore
" ontogeny " briefly recapitulates " phylogeny," the individual organism passing rapidly through the stages
by which evolutionists hold that the species has grown
to be what it now is. The multicellular or higher
organisms are only, as it were, loose appendages to certain peculiarly qualified unicellular organisms, like great
flickering shadows of dwarfs or little children cast by a
bonfire. The higher organisms perpetuate themselves
qua unicellular. They may seem bicellular, because of
the curious sexual split into male and female ; but we
must remember that ovum and spermatozoon combine
in one to form a new life-history. And all the future
of the individual life lies *in nuce* in that single cell.
And we can further trace this determination of the
qualities of maturity by the qualities of the embryo
right back through the continuous germ plasm to an
age when the whole world of organisms was unicellular.
No fresh quality has come to any living creature since
life began its ascent. All were implicitly present in
the unicellular world ; all have been slowly evolved and
improved by nature's gigantic game of permutations
and combinations. She has written out by degrees every
possible grouping of the qualities of protoplasm, and has
drawn her pen remorselessly through the inefficient ones.
The favourite image or parable for this view of heredity
—given, *e.g.*, by Huxley in the notes to his Romanes
lecture—is that of a plant propagating itself by suckers.
Root grows from root ; every here and there the root
sends upwards a perfect plant, a glory of leaves, flowers,
fruit ; in the absence of these the root could not be
healthy ; yet plant is never derived from plant, and
still less is root derived from plant ; every root is
derived from root ; every plant is derived from root.

Another image we might use is that of a river like the Nile, flowing through countries which can yield it no tributaries. The great river flows majestically on, essentially the same as it was many hundreds of miles up channel; imparting life wherever it goes, but receiving nothing. Such a river of life is "germ plasm," flowing through the generations, yielding to all of them support, but never affected by them.

There is, however, a difference which our images fail to bring out. On Weismann's view, evolutionary change is always at work, acting through natural selection. Permutations and combinations are always being remodelled—let us say, combinations of playing-cards. The cards were originally dealt at the dawn of animal and vegetable life; and no fresh kind of card has ever been introduced. Yet the "hands" with which the game is played have, on the whole, steadily improved from generation to generation, and from age to age. How is that possible? Because these cards are alive. These cards multiply, aces begetting aces, and kings begetting kings. Many and many a hand has been torn up and flung away in the process of natural selection; and accordingly the surviving hands have become very strong — all court cards, or trumps, or powerful sequences.[1]

At the back of this process of combinations we have another—the original dealing or the original making of the cards. To what was that due? To the Lamarckian factor, to the direct action of environment, stamping itself upon the isolated living cell. There is an absolute contrast, it is assumed, between the two periods in

---

[1] I do not know if Weismann means this; but it seems to lie in the theory. Efficient begets efficient, as surely as non-efficient begets non-efficient. *Quantities* seem capable of indefinite improvement, though the theory admits of no fresh ultimate *quality*.

the history of life. In the first, variations were due directly to the environment, not at all to natural selection,[1] which only acts upon variations submitted to it by sexual reproduction. In other words, environment may be called the judge in natural selection, but there is no need of environment as a judge when it is itself the maker of the things to be judged. If it is the maker, it gives a guarantee along with its goods. If or so far as Lamarckism is true, Darwinism, with its "natural selection," becomes secondary if not superfluous, ranking at best as an auxiliary and accelerating force. Thus, if the unicellular organism bears the stamp of environment, it has directly adjusted itself to the conditions of life; it is already certified as "fit to survive." But, in the second great period, we are to believe that environment is helpless and natural selection omnipotent. This is less arbitrary than it seems. In the unicellular age the living creature is all surface, and, as it were, at the mercy of environment. But in the multicellular age the really vital matter, the "germ plasm," is supposed to be carefully hidden away inside a body and out of reach—hidden within a body and even (the theory says) independent of its vicissitudes, so long as the body lives. The only way in which nature can now affect germ plasm is by killing off the body in which it resides, under sentence of "unfitness." Thus indirectly — natural selection is always indirect — and slowly—indirect processes of course are slow—evolution is pushed on. For in this fashion germ plasm is progressively improved; and unicelullar embryos, needing nothing from the mother beyond nourishment[2] up

---

[1] So Weismann as stated by Romanes.

[2] Heredity is equal from the two parents. It seems therefore that Weismannism must be right in denying that the fœtus draws anything beyond nourishment from the mother organism.

to and after birth, come to contain in themselves the
promise and potency of reason, of genius, of greatness—
of a Shakespeare or a Darwin.   A little speck of matter,
indistinguishable to human study from one of the
lowest forms of life, and essentially nothing but one
of these lowest forms, redistributed or regrouped, contains in itself what will necessarily ("bar" the accident
of death) give the difference of a man from a beast, of
a genius from a fool, of a saint from a scoundrel, or *vice
versa.*   So runs the doctrine.

We have not yet stated Weismann's ingenious
theory that the germ plasm, and unicellular organisms
in general, are potentially immortal.   Unicellular organisms grow by fission ; the child is a part of the
parent; it is impossible to say, after the split has
been accomplished, which is child and which is parent.
Both are both ; or neither is neither !   The category
or conception of parentage belongs to a higher sphere
of life, and is inapplicable here.   If either survives—
and we are assuming the continuance of the species—
both may survive.   Each member of the race is potentially immortal.   Never a natural death, but a violent
death always, must weed its ranks.   If germ plasm
exists at all in continuity, it is hardly necessary to
argue that the same thing must be true of it.   Part of
the germ plasm builds up a body, and undergoes in
somatic form the doom of death ; part of the germ
plasm survives as germ plasm, multiplying and re-
plenishing itself (if only during embryonic growth),
and ultimately—in some fortunate fragments—passing
into new lives.   This thing need never die.   Most of
it will die ; what is transformed into body, and what
fails of attaining to fertilisation.   But it need not die ;
it is potentially immortal.   So to say, the old original

germ plasm may *hand on* the duty of building up a body to some of the more newly formed material, and, evading the chances of death, may refuse to quit the parental tissues till the moment of fertile sexual inter-course.   It is potentially immortal; practically, by the law of chances, it will be both mortal and short-lived.   If pollen grains depend on the wind or depend on insects for doing their work, how much potentially immortal "germ plasm" must die in the history of every diœcious plant!

Unicellular creatures, however, are immortal, accord-ing to Weismann, rather *qua* non-sexual than *qua* unicellular.   Sex and death are somehow correlated; he believes that he has proved this by showing a general correspondence between the age at which species pro-duce offspring and their natural term of life.   This view of Weismann's is widely accepted.   A correlation be-tween the fact of sex and the habit of dying a natural death is largely admitted.

Death, then, as a natural and certain event, arose with sex, or in consequence of it.   But how did sex originate?   Romanes asserts a self-contradiction in Weismann, because at one time he says that the origin of sex was due to natural selection, at another time that it could not be.   In Weismann's system, natural selec-tion works upon the materials furnished to it by sexual reproduction—upon the new varieties thus invented—upon the new permutations or combinations of germ plasm, thus manifested, and brought up for judgment in the form of offspring.   Still, I see no reason why natural selection should not sit in judgment upon sex itself, if sex somehow originated.   No doubt the admis-sion must then be made that Weismann's clear theory of variation had ceased to be available.   Sex explains other

variations ; what is to explain sex ?   It must presumably
itself have been a new variation when it appeared for
the first time in a sexless world.   Once it had appeared,
it might well predominate.   If some multicellular organisms propagated sexually, and others non-sexually, and
if some of the offspring of sexual unions proved superior
in the struggle to any of their competitors, why then
sex would be *selected* by *nature*[1] as advantageous ; the
sexual specimens would *tend* to be the only ones that
survived and reproduced their kind.   The origin of
sex, accordingly, would still be veiled in deep darkness.   Weismann could say little more than that it
"happened" to occur.   That is very much what he
does confine himself to saying "in the present state of
our knowledge."   Yet it appears perfectly logical to say,
not that natural selection brought sex into being;
*natural selection originates nothing; it chooses be-
tween competing candidates;* but that, from the first
and until now, natural selection has favoured sex, and
has made it the predominant reproductive method.
This seems to be perfectly fair, if Weismann is willing .
to postulate the true condition of natural selection,
viz. competition ; in this case, competition between
sexual and non-sexual forms.   But I am afraid that
may not be so.   In view of Weismann's attitude
towards the question of the origin of (natural) death,[2]
one must concur in Romanes' criticism, that "ultra
Darwinians use the term 'natural selection' with
extreme laxity."   The condemnation might be even
more severely expressed.

[1] May we say that, upon the whole, it is selected by nature, at least
for the higher forms of life ?

[2] See the paragraphs which follow.   Of course, if there is a correlation of sex and death, the new question is really the same under a
different name.

As to the origin of death, I must confess to finding the theory most unsatisfactory. Of course we are speaking of the origin of the habit of dying a natural death. Death by accident, death as prey, death (possibly?) by disease, may all be assumed, independently of this new and advantageous habit of retiring the seniors at a (roughly) fixed period. The new habit is said by Romanes to be advantageous for this reason, because, if multicellular (or, as he says, if sexual) organisms lived through ages, they would all become broken down and decrepit as the result of accident. For the life of me, I cannot see why this should be true. If there was any emergency with which unaided natural selection was able to deal, I should have said it was this one imaginary danger. Will the poor old things not be overtaken by their enemies? Will they not starve from their prey escaping them, or being taken from them by younger competing creatures of their species who are *not* run down by accident or infirmity? Have we any reason to believe that natural death—death from old age—has ever been common in the animal world (in plants, perhaps, yes); or have we any reason to regret its absence? But, if it plays a scanty part, how could it secure the attention or obtain the approval of selecting nature?

Next let us ask, how we can conceive of the process of selection being accomplished? Race A is competing against race B. The prize is fitness to *survive;* the penalty, of course, is just death. But race A, being clever enough to invent the habit of dying a natural death, *therefore survives*, while race B, which refuses to die unless by force, is therefore extinguished.

This is not altogether such an Irish bull as it sounds.

It may be held that a habit in any species of dying a natural death will produce a more efficient average individual. And so it might be possible, given the conditions, to think out the mechanism of the process. Here also, of course, natural selection does not originate the habit in question; in this case dying. Death may be, as Weismann seems to hint, in obscure physiological correlation with the conditions of sexual reproduction. It may be put down as a "chance," *i.e.* until now an unexplained, variation.[1]   One race "happened" to begin dying off, and profited thereby *qua* race. From it sprang all the winning species, or else the same thing "happened" over and over again. Death might also be said to be involved in certain permutations and combinations of the germ plasm. That is the beauty of this unknown and unknowable substance. Nobody can say what it may not imply. If a rearranged protozoon implies a Beethoven or a Shakespeare, if it gives him his programme, "Be thou among the greatest of the sons of men," molecular rearrangement in a germ cell may well imply the simpler programme, "Thou shalt surely die." And so, if he likes, Weismann may claim this memorable "variation," natural death, as due to the cause by which he seeks to explain the origin of all variations.

That, however, is not Weismann's line. Instead of that he protests that, in calling natural selection the cause of death, he does not mean to imply any competition between naturally mortal and potentially immortal stocks. Then pray what right has he to talk of natural selection? Let us go back to first principles. How does Darwin's title-page define natural selection? As

---

[1] Use-inheritance will do nothing here. A habit of dying, after it has been acquired, assuredly cannot be transmitted to offspring!

"*the preservation of favoured races in the struggle for life.*" If there is no struggle for life, and no preservation of a favoured race, neither is there any natural selection. Weismann's usage is worse than "extreme laxity." It aims at finding something cabalistic in natural selection, something talismanic. He must be reminded that, according to Comte, "nature" is the supreme example of an empty abstraction by which "metaphysical" persons think to explain phenomena, while giving no explanation at all. Weismann is a "metaphysician" of that type. He uses the phrase in lieu of an explanation, not knowing, and not caring to know, what he means by it.

In taking leave of Weismann's fairy tale, it may be desirable to name one by one his characteristic positions, and to add in regard to each whether he still retained it in 1893, or had modified it, or had cancelled it.

First, Weismann used to hold that protozoa and protophyta—unicellular nucleated plants and animals, the lowest forms of life known to us—were exempt from natural selection, and were subject to the agency of environment as a source of variations. Convinced by the experiments and arguments of other writers that conjugation and natural selection were both at work in these creatures, he has come to postulate still simpler forms of life unknown to observation—creatures without even a nucleus—creatures (though not the only creatures) which are potentially immortal. Now, it is an immense weakness to have to postulate unknown forms of the living organism. Yet perhaps it may be contended that this one addition to the theory is sufficiently logical and coherent. Even in the protozoa and protophyta, as an unscientific person might say, "germ plasm" is hidden away in a nucleus, if not behind the wall of a special cell. In purely homogeneous living organisms, if such existed, all parts must share and share alike in the interactions between organism and environment.

Weismann held that the protozoa and protophyta were potentially immortal; also the germ plasm. All these positions stand, or stood, up to the date of 1893.

He held that sex originated in the course of evolution, and was absolutely due, whatever that might mean, to *natural selection*. This he still maintained.

A similar view had been broached by him as to natural death; he still maintained it.

He had formulated a doctrine of "germ plasm." This has been modified, refined, elaborated, re-christened, and, in fact, transformed more than once, both before and after 1893. But this and other technical changes of great importance do not sensibly affect the "fairy tale," nor the basis of Mr. Benjamin Kidd's social gospel, preached by him in the name of Weismann. We do not therefore dwell upon these changes.

Next there is a group of three very important points, which imply each other, and stand or fall together; that amphimixis is the only cause of variations; that environment is impotent to originate them, in view of the "continuity" and "absolute stability" of the germ plasm; that every higher and highest organism is simply a unicellular organism of an improved or rearranged kind, with its appendages and necessary consequences. The central point here is the stability of the germ plasm. *Weismann gives that up* (1893). The second point of our present group of three is therefore gone. In consequence the first point must be at least modified, and it turns out to be absolutely inverted. Amphimixis is *never* to be the cause of variations; they are to go back to differences and irregularities in nutrition. At the same time, by a curious codicil, Weismann insists that these differences could never become effective unless when they were cumulated by amphimixis. According to Romanes, this is simply a piece of obstinacy designed to show that, if Weismann was certainly half in the wrong, yet he may have been half in the right. Romanes therefore thinks it is to be dismissed as an unproved and improbable assumption. The third point also falls to the ground. The germ plasm of one of the higher plants or animals or men is not simply a one-celled creature rearranged; it is such a creature, if you like, but modified as well as rearranged — modified to a certain extent all along the course of its "phylogeny," wherever variation occurred.

Modified how far? That is for us a very important question. Do Weismann's newer views admit of use-inheritance in the literal sense? Or do they only admit of certain changes in the germ plasm, sympathetic to vital changes in the parental organism, but not necessarily initiating the same changes in the offspring? In Romanes' language, does Weismann now accept *representative* congenital changes (= true use-inheritance), or only the lower class or classes, nutritive changes (= Weismann's new theory of the origin of variation), or nutritive and specialised?[1] This is a question of importance for us as students of human progress. True use-inheritance, if it occurs, constitutes a possibility of rapid advance in contrast to the painfully circuitous method of natural selection. So far as I am aware, Weismann has not spoken on this point. Reluctantly, and as it were casually, he has cancelled the central

---

[1] Romanes gives as an example of the last: "The fathers have eaten sour grapes, and the children" were born with wry-necks!

doctrine of his scheme, that of the absolute continuity
and stability of germ plasm.   It must be deemed at
least possible, according to Weismann's later views, that
use-inheritance should take place.   The question will
demand more imperiously than ever the eliciting of an
answer from facts.   Accordingly, when Mr. Benjamin
Kidd builds his sociology on the absolute non-inheritance
of acquired qualities, he is building on a rock perhaps,
but on a rock whose discoverer himself has undermined
it and stored it with explosives.   This is not our only
objection to Mr. Kidd's premises, but even in itself it is
a grave matter.

It is possible to postpone as a merely technical point
the question, whence come the variations with which
natural selection deals ?   So long as such variations *do*
arise, it may be said, there is little need to trouble ourselves with the how or the whence.   But Weismann's
dealing with the question is less vigorous and rigorous
than it was.   His fairy tale has suffered.   As they now
stand, his doctrines are less astonishing, and somewhat
less incredible.

There is still one more point to name ; we may call
it the second basis of Mr. Kidd's sociology.   It is held
that where progress ceases you have in its place not
stagnation, but actual retrogression.   No progress, but
by natural selection ; nothing but retrogression, where
panmixia prevails.   So far as I am aware, Weismann
has never recanted this position,[1] which has tremendous sociological consequences in Mr. Kidd's hands.
Yet it seems a characteristic bit of the newest science,
a piece of purely deductive reasoning from facts, or

---

[1] In 1895 he made the admission that panmixia could not in itself
fully account for retrogression, though it tended that way ; and the
obscure doctrine of germinal selection was brought in as a supplement to
panmixia.

from a mixture of facts and theories, and a deduction of doubtful logical coherence. Scientific friends inform me that there is great division of opinion among men of authority [1] on the question how panmixia must work out. Will it mean continuous retrogression? Will it reach an average mediocrity and stop there? Will it mean a divergence into two or more distinct types? Doctors differ. Surely then Mr. Kidd has planted his feet on a second slippery stone. As a matter of obvious probabilities one does not see how continuous embodiment of the stable germ plasm of to-day should or could mean continuous degeneration and progressive inefficiency. On a first glance, at any rate, that view seems absurd. And the division of opinion among biological experts emboldens one to break away from the dogmatism of Professor Weismann and Mr. Kidd.

[1] Professor Baldwin, the psychologist, refuses, for one, to admit Weismann's theory of necessary retrogression.

CHAPTER XIX

HYPER-DARWINISM IN SOCIOLOGY : STRUGGLE MADE
ABSOLUTE—MR. KIDD

Resemblance to Comte—Intenser emphasis on biology [cf. Mr. Platt-Ball]—(1) Panmixia = degeneration is inconsistent with dreams of socialism or of final balance—Selfishness, however, may not care for remote consequences —[Ought Panmixia further to imply extinction ?]—Also, social "statics" are blotted out—And evolution becomes almost identical with progress— Could not Mr. Kidd save many essential positions without this assumption ? —(2) Next, if progress implies struggle—And selfish reason makes un- willing to struggle for good of the race, supernatural counterpoise of religion must help, as hitherto—Now, Weismann had riddled his own position with qualifications—Kidd also appeals to biology by a doctrine of the social organism ; but everything here depends on philosophy, not biology—(3) First, the doctrine of reason ; reason is formal, as with A. J. Balfour, Darwin, Drummond—For Mr. Kidd also holds that biological law applies without a break to rational man—Yet reason *disturbs* process of evolution—And Bagehot, Stephen, Drummond have noted other changes due to it—Can it be wholly evil ?—Balfour and Kidd repudiate Kant or Coleridge's deeper sense of "reason"—But they cannot avoid such sense if it lies in the word and in the fact—(4) Secondly, doctrine of religion as anti-rational—Not = "future judgment" ; that is rational !—Can we believe the irrational ?—Does not Kidd tamper with Christian equali- tarianism ?—Biologically ; variation *may* be purposeful and professive— Historically ; reason *is* progressive ; by rational methods—Religion its fulfilment—It needs a force to give it motive and constancy.

THERE is a great deal to recall Comte in *Social Evolu- tion.* We have a long and interesting appeal to history. We have the doctrine of altruism assumed, without inquiry or justification, as a definition of the moral ideal ; though it is ousted from the place of legitimate authority which Comte gave it by Mr. Kidd's anarchical conception of reason as purely selfish, and has to borrow its credentials from non-rational religion.   Above all we have the appeal to biology more unhesitating than ever.[1]   "It may be remarked that nothing tends to exhibit more strikingly the extent to which the study of our social phenomena must in future be based on the biological sciences, than the fact that the technical controversy now being waged by biologists as to the transmission or non-transmission to offspring of qualities acquired during the lifetime of the parent, is one which, if decided in the latter sense, must produce the most revolutionary effect throughout the whole domain of social and political philosophy."[2]   Yes, it is striking ; most extremely striking ; so remarkably striking, indeed, that one would have expected the author to reconsider the question, whether it is necessarily true, if not to raise the question, whether it is even possibly true. Comte himself, phenomenalist to the backbone, while insisting on the connection of sociology with the lower science of biology, insisted also on its separate province and independent laws.   Now it appears that sociology— like one of the colonies of France—is to be merged outright in the mother empire.   Everything is to be biological.   Human wisdom, for the most part, is to be an incidental deduction from the laws of life, as manifested in four-footed beasts and fowls and creeping things of the earth.   Is it really the case that the progress of science since Comte makes this conclusion inevitable ?   Or is it rather a retrogression in the

---

[1] Page 203, towards end of chap. vii.   The same thing is to be noted in Mr. Platt-Ball's little book against use-inheritance (see Preface, p. vii.)

[2] Mr. Kidd differs from Mr. Sutherland—(1) in appealing to the working of *struggle* rather than that of *elimination* among mankind. Neither really succeeds in appealing to the struggle, or to the elimination, implied in true natural selection ; (2) Mr. Kidd allows reason to do something—it makes mischief !

higher culture — a relapse from the not too lofty philosophical sympathies of Comte — which gives us the proposed biological tyranny? It is an excellent thing, that each man should be an enthusiast for his own speciality; assuredly it is an excellent and healthy thing; but there are limits!

The doctrine of inevitable retrogression when progress ceases — which we noted in the previous chapter as Mr. Kidd's second great debt to Weismann — has important consequences for sociology. It sweeps away socialistic dreams, as well as Spencer's doctrine of a stationary state. The second will probably find few mourners to shed a tear over it, though it may be difficult to give up the purely *economic* conception of a stationary state. What will happen when the world is absolutely too full, and population must cease to grow? That is one of the unrevealed mysteries of Mr. Kidd's *credo.* Will he tell us the world is not going to last so long? Will he appeal to a struggle for eminence as doing the work of the old struggle for life? In the latter case much of his book would need reconsidering. As to socialism, he points out with much force that arguments which show it to be unscientific may yet fail to dislodge it from the minds of men. Sociological science warns the socialists, " You will retrograde,[1] and

---

[1] There are two points here : (1) you will retrograde, because natural selection will cease ; (2) natural selection will extinguish you, because you have retrograded. The second will only hold true if socialism and stationariness are partial. Like the eight hours' movement, or like bimetallism, socialism (etc.) must seek to come in by international arrangement if it is not to be speedily swamped by competition from hardier races, *within* which natural selection is still going on. But, if it were an international possibility, the whole world might jog quietly down hill (see p. 315).—That is the theory. Facts do not seem as yet fully to bear it out. France is still a great power, though perhaps in a perilous way (Feb. 1899). And at least France is being swamped by the more prolific races.

therefore your posterity will soon be extinguished."
Suppose the socialist to reply, " What on earth do I
care about posterity ?   I mean to have an easy time of
it myself!"   Then certainly your remonstrance has
missed fire.

Another consequence of some importance for sociological science attaches to this second great loan of
Mr. Kidd's from Weismann.   The old Comtist and
post-Comtist division into statics and dynamics—conditions of order[1] and conditions of progress—falls to the
ground.   Mr. Kidd discusses the "conditions of progress," and these only.   The formula seems to be,
"Take care of progress, and stability will take care of
itself"; a formula which follows directly from Weismann's dilemma—advance or downright retrogression—
and yet once more so startling a position that once more
it seems Mr. Kidd ought to have been arrested, as
by a large type note of interrogation or by a danger
signal, and ought to have inquired whether something
had not been ignored when biology was transferred
wholesale to the life and history of man.   The young
lions of the Radical party will welcome Mr. Kidd's
formula with delight ; but one would rather hear what
the old lions have to say to it.

Yet another consequence may be noted ; evolution,
with Weismann and Mr. Kidd, is almost though not
altogether equivalent to progress.   It is progress where-
ever it is not downright retrogression.   Stagnation is
impossible, panmixia and retrogression are rare.   No
doubt panmixia will yield continuous evolutionary
change while it lasts ; but panmixia is essentially a
limited phenomenon ; it is an exception to the general

---

[1] Comte's *Statics*, however, as he states them are rather *abstract conditions of social well-being* than conditions of social *order*.

rule. It may prevail in solitary islands, literal or metaphorical; but the great tides and continents of life are peopled by struggling, suffering, progressive creatures. On a broad view, evolution means progress.

Before leaving this assumption, it may be well to ask how much depends upon it? Go on, or you will go back; acquiesce in struggle, if you don't wish to retrograde; that is a very urgent appeal—an overwhelming appeal, one might call it. Yet in many respects the same result might be reached by the narrower and less urgent, yet tolerably effective appeal, acquiesce in struggle if you wish to progress and to avoid stagnation. Few of us would be content with a "stationary state" from the present hour and onwards. The narrower appeal would hold us. The same practical results would be reached, with a less precarious and less vulnerable array of assumptions. Socialism would still be condemned as arresting the further progress of the species. Evolution and progress might still be regarded as equivalents—perhaps more so than ever; but we could reopen the book of Social Statics, and admit (for those who desired it, or who felt bound to anticipate it) visions of an ultimate stationary state.

We pass now to Mr. Kidd's first basis, assumed from Weismann, the doctrine that all progress implies struggle and natural selection. This doctrine yields the first or almost the first abstract formula for social dynamics. Comte and others gave us historical sketches and sequences, not general principles or causes of progress.[1]

---

[1] If Comte had formulated these, they might have found their way into his Statics rather than his Dynamics.

What then are the conditions of human progress as formulated by Mr. Kidd? Primarily they are physiological. Let men fight the battle of life; they will advance. Easy circumstances, enjoyed in an easy spirit, imply arrest, and perhaps arrest implies retrogression. But the wholesome biological tendency to struggle, and struggle on, is interfered with by man's gift of reason. The instincts of race keep the beasts in the path of progress, *e.g.* by struggling in the interests of their offspring. But many human beings— *e.g.* the school of Mrs. Mona Caird—resent these struggles as an impertinence and an absurdity. So far, Mr. Kidd agrees with them. It *is* irrational to acquiesce! Reason makes us conscious of self; selfishness therefore and selfishness alone is rational behaviour. But rational behaviour, in this sense of the word, leads straight to retrogression. Now, natural selection would have its slow remedy for this. If the human race had entered the *cul-de-sac* of selfishness, natural selection would calmly have waited till a rational race endowed with higher tendencies "happened" to be evolved; whereupon humanity would quickly have been extinguished in competition with the new race. But, fortunately for the prospects of mankind, such an evolution has already "happened." Mankind *is* a race fitted to survive. Or rather—Mr. Kidd does not write on this point like an ultra-Darwinian, giving the largest possible play to chance, but like one who has a belief in the purposefulness of organic life — the biological laws of human society supply a counterpoise to the dangers introduced by reason. We have reason to make us selfish; but we have religion to make and keep us altruistic, in despite of our reason. All religions are preter-rational and altruistic, Christianity the most of all. So we have been swayed, and have struggled, and have progressed. We have struggled in war. We have struggled by mere contact, when lower races have melted away at the presence of civilised man. We are struggling to some purpose in the scramble for Africa. And, beyond a doubt, we of the white races shall succeed in a further struggle to control the Yellow Terror for the greater good of humanity. Mr. Charles Pearson's formidable table, proving the rapid increase of blacks in the United States, is met by another set of tables, proving that the increase is not so rapid after all. Such effect has Christianity had in making us altruistic that we have voluntarily widened the sphere of rights within each nation. Yet we are not drifting into socialism. Quite otherwise ; what the *Zeitgeist* really means is to secure genuine equality of opportunity—intensified struggle between citizen and citizen —accelerated progress ! The yielding of Militarism to Industrialism, and the allied change "from status to contract," are earlier stages in the same great development by which competition grows ever more and more intense.

Here then we have Comte's three appeals brought into odd harmony with an apology for supernatural or at least for "ultra-rational" religion. This is to be heartily welcomed as an advance in the right direction ; and the criticisms passed by Mr. Kidd, on the contemptuous treatment of the origin of religion by Mr. Spencer and his underlings, are well deserved and well established. A saner view of history certainly does commend the opinion, so powerfully advocated by Seeley, that religion is the great animating force in states and societies, the master-builder of historic greatness. Nor can it be denied that there was need of reaction from a one-sided intellectualism, which had prevailed even in quarters [1] where we find but little faith in reason.

Granting all this, and granting it gladly, one must go on to express grave distrust of the process by which Mr. Kidd reaches his conclusion ; of the terms in which he formulates it ; and of the affirmations with which it is connected.

First, even if we accepted the claim that biology was to be the final judge, we must regard Mr. Kidd's Weismannism as a very insecure foundation. We have already noted in some detail how the denial of use-inheritance had been qualified and weakened and transformed by its author even before Mr. Kidd applied to Weismannism for a social gospel. And we have seen that the doctrine of necessary regress in the absence of struggle and consequent advance is a precarious deduction from Weismann's own premises, and is scarcely necessary to Mr. Kidd's sociological system.

Hitherto, however, we have considered only one form of Mr. Kidd's dependence on biology.[2] So far, we have spoken of his doctrine concerning men *qua* physical organisms, exposed to the same conditions as other living creatures. A different use of language by Mr. Kidd must now be considered. His further doctrines regarding reason and religion are brought into connection with biology by means of the familiar phrase, the social organism. True, Mr. Kidd thinks that other writers who have used this phrase have led us very little, if at all, further on. Still, it points us in the right direction, and the new guide is confident of securing better results.

---

[1] *e.g.* Mill and Buckle. See below, in the closing paragraphs.

[2] Professor Lloyd Morgan shows very tellingly that Mr. Kidd is not warranted by any facts he adduces in contrasting man's intellectual and his moral evolution (*Habit and Instinct*, p. 345). Yet another part of the case therefore breaks down.

Not man the individual, but society as such is now viewed as illustrating biological law. There are conditions of vitality or of progress—progress is a manifest fact; there are difficulties revealed by observation or by consciousness; and there are safeguards or remedies discovered by analysis. This does not sound very like Darwinism, still less like Weismannism, though it is brought forward as based on the latter. The truth is, the basis here is nothing; "social organism" is only a phrase; the analysis here is everything. All depends upon the truth or erroneousness, the worthlessness or the value of Mr. Kidd's doctrines of religion and reason. In dealing with these points, he must speak as a philosopher. His biological knowledge does nothing here to guard him against error.

The doctrine of reason is similar to what we find in Mr. A. J. Balfour's *Foundations of Belief.* Each writer, in a footnote,[1] repudiates any higher or deeper doctrine of reason than that which regards it as a calculating machine or process of inference. This implies that reason is passive in knowledge, and plays no part in determining the motives of human conduct. The effect of the latter belief, when held by intuitionalists, is that they postulate a moral faculty of conscience alongside of reason and independent of it. In Darwin the effect is this, that moral motives are interpreted by the animal impulses of gregarious creatures, impulses which are held to be extended in range, but not altered in quality, by the advent of reason. And in Drummond the effect is that he looks for one set of impulses which even in animals may be labelled good and right, in

---

[1] *Social Evolution*, p. 73, 2nd edition. *Foundation of Belief*, p. 195.

contrast to mere self-seeking.   Only by such a discovery
is Drummond able to save morality.

In assuming that biological law may be applied *en
masse* to human conditions, Mr. Kidd seems to re-
affirm the doctrine, that reason has no material influence
upon motive.   Yet it turns out otherwise.   He does
believe that the animal nature of man is affected by
reason, viz. for the worse!   Conscious of what he is
doing, man objects to sacrifice himself to his family or
his tribe; instinct might have led the ape to make the
sacrifice automatically.   Reason thus tends to make
man purely selfish; and sometimes the tendency has
its full effect.   After all, selfishness is the only reasonable behaviour.   If indeed reason can be controlled,
it promises great social advance through the superior
cleverness which it imparts; but in itself it is a purely
anarchical force.   De Maistre or Newman could not
have spoken more severely of it.

Let us recall here what we have learned from other
evolutionists regarding the advent of reason.   It has
arrested the evolution of the body (Drummond, etc.).
It has wrapped mankind round in a mantle of law,
custom, and institution, capable of intellectual not
physical inheritance (*e.g.* Mr. L. Stephen).   It has
largely substituted imitation or conversion for rivalry to
the death (Bagehot).   And now Mr. Kidd tells us that
reason abruptly closes—so far as its influence extends
—the process of upward social evolution.   Does not all
this support the conclusion that reason is something
quite different from a mere colourless medium or
calculating machine?   One fully agrees with Mr. Kidd
that reason checks the automatic working of instinct.
Where reason appears, systematic selfishness and sin
become possible as they never were before.   But

unselfishness too becomes possible as it never was before; it has a new significance. Reason has broken up the unity of the life of sense. Does it do nothing except break it up? At the lowest, is reason not shrewd enough to perceive the unhappiness of a selfish life, the greater gain to oneself of a life animated by unselfish and far-reaching interests?

Something must be added here regarding the use of the word reason by Mr. Kidd and Mr. Balfour. Reason is narrowed by them to reasoning, and even (*pace* Mr. Balfour) to rationalism. Mr. Balfour's footnote seems to be dealing Coleridge a sly hit when it repudiates acquaintance with the Logos. Now no doubt Coleridge had a provoking habit of exclaiming "Logos" as if it were a talisman of magic power. We have seen something similar in our own day on the part of that very able and powerful and now venerable Hegelian writer, Dr. Hutchison Stirling. In his case, "the Notion" was the talismanic word. Mr. Kidd again goes straight to Kant,[1] by whom, of course, Coleridge was influenced. But Kant is very obscure. Some provocation had then been offered the plain Briton. And the way in which the doctrine of Reason or Logos shaped itself with Kant or with Coleridge— in many points alike; in many points, also, not alike —was open to further criticism. Every doctrine of "faculties" is, to a large extent, artificial. Reason and Understanding shade into each other, however we may choose to contrast them.

But, just on that account, the plain Englishman will find it hard to keep clear of the deeper and more

---

[1] Without reporting him very accurately. Grave objection might be taken to the formulation of each of the three great Kantian positions given by Mr. Kidd.

mystic features of reason. He wants to be a practitioner in the simpler branch of the art ; well ! the arts are not two but one. His own words will prove disobedient to him. Words are something more than the clothes of thought ; they are its incarnation. We inherit words ; we use them in our service, ennobling them or, more frequently, debasing them ; they lived before us, and they will long outlive our very memory. We are the fleeting shadows ; they are the substances. Words are like homing pigeons ; they will carry our messages, if we manage them wisely ; but with an instinct surer than our choice—with an instinct not to be overborne by our caprice—they will go *there*, to that one point where each is at rest. If we take up the great task of the impersonal reason of mankind, it is in vain that we express our determination to keep clear of the transcendental or of the logos ! It is in us and we are in it ; in it, or in Him, we live and move and have our being, unless Mr. Balfour carries us off in his alluring company upon one of his favourite excursions to " a standpoint *outside of* reason." Inmates of a madhouse are as nearly as possible emancipated from the logos ; to all others the logos is " closer than breathing."

Mr. Kidd's doctrine of religion is largely determined by his doctrine of reason. Reason, though useful (like fire) as a servant, is, like fire, a thing anarchical and destructive. Religion, the source of order, is, by the very nature of the case, extra-rational. Religion makes it man's interest or man's impulse to do things which are not personally for his profit, and which reason therefore discourages.

At first blush, one is tempted to connect Mr. Kidd's doctrine of religion with the familiar doctrine of future rewards and punishments. These are represented as supernatural motives for doing good. They are not, however, extra-rational; they make it worth one's while to be moral. Righteousness is strictly a rational and self-interested policy, if this be the truth. This is not therefore Mr. Kidd's meaning; and the doctrine in itself is unsatisfactory. Selfishness produced to infinity remains selfishness still; it does not turn into righteousness or unselfishness. Other worldliness is only a more morbid growth from the same root as worldliness. If it is moral—if it is one's duty—to preach the doctrine of future judgment, that is only because selfish fears and selfish hopes, once awakened, may be transformed, without a visible break, into something nobler than themselves. They are moral protoplasm (in the true and Aristotelian sense). They are the germ, though only the germ, of goodness.

When once, however, we have shut out this interpretation of Mr. Kidd's doctrine of religion, it is very hard indeed to say what the doctrine means. Religion works powerfully, but irrationally; that is all we are told. It sounds as if religion were a sort of white magic or hypnotic influence. It sounds like a revival of opinions held by wise men under the Roman Empire, according to Gibbon, when all religions " were considered by the people as equally true, by the philosopher as equally false, *and by the magistrate as equally useful.*" Religion serves the public weal; religion augments altruism; the Christian religion in particular attains its ends by a sweeping dogma of human equality. But how Christianity or any other religion captures the wills of human beings, of that we have no explanation. And when we find that Mr. Kidd, in view of the scramble for Africa, and of the

taking of the black races under white tutelage, thinks
that Christianity must consent to modify its equali-
tarian dogma, a dogma that has been so operative
and so useful in the past, one surmises that a high
appreciation of the past usefulness of the Christian
religion is quite compatible with a very cool and
detached consideration of its claim to present authority.
Indeed, can any man believe that which by definition
is non-rational ?   And—to take another point—is not
Mr. Kidd's proposed tampering with the rigour of
Christianity a most unholy piece of rationalism ?
Alas !   The countrymen of Cecil Rhodes. seem in small
danger of being irrationally altruistic, or democratic, or
humanitarian in their treatment of the black man !
And, if the premises are true, is not Mr. Kidd's personal counsel most subversive and pernicious ?   If religion blindly obeyed in the past has made us what we
now are, must we not still obey religion with what is
called blind fidelity ?   If irreligion has brought its
penalties hitherto, will not irreligious acts incur the
same doom hereafter ?   And irreligious theory no less !

Biologically, Mr. Kidd seems to have left one
possibility unconsidered.   Congenital variations may be
due to the environment (by use-inheritance or by
differences of nutrition), or they may be due to
amphimixis ; or thirdly, they may be due to an inner
tendency to vary.   Mr. Kidd, in his enthusiastic
adherence to Weismann, has left the last possibility
out of consideration ; yet Romanes points out that
Darwin was inclined to look in that direction.   Now, if
there is a tendency to variation in living species, if
variation is not simply forced on them by environment,
there is no reason for assuming that variation will be

purely casual or non-telic. The embryo is a wise little architect, who builds up a new life out of a speck of protoplasm by the help of nutritive materials. He makes no mistakes; he gives us new organisms each after its kind, each perfect in every part, unless where mere force damages his work. If this wise little architect varies his plan slightly, it is far from being obvious that he varies at random. If he knows so much as he plainly does know, should we not give him credit for knowing a little more? If he knows enough to keep him faithful to the plan of the specific type, ought we not to believe that, when he introduces variations, he knows what he is doing, that he makes improvements, not random shots? He is not a piece of lifeless mechanism. He is a standing miracle — a " natural supernatural." We are confidently told that the abandonment of belief in preformation and adoption of the theory of epigenesis was a heavy blow to teleological and theistic doctrines. I confess I should have thought the opposite. Is there not more of the likeness of miracle in the emergence of an organism, true to its own type, from a speck of living jelly, than in the growth of a detailed miniature by mere accretion in bulk? Be that as it may, there is at any rate no literal preformation, and there is the fulfilment of purpose. Then, if variation occurs spontaneously—from the resident forces of life itself—can variation be a thing of random direction?

Now random variations may become purposeful if they are well weeded by natural selection. But variations which are purposeful from their very beginning— like those due to use-inheritance, if such are really transmitted—do not need to be sifted by the elaborate and tedious process of natural selection. It is perfectly

conceivable that purposeful variations[1] occur spontaneously in each species, and are a direct source of
progress.

When we leave biology for sociology and the sphere
of reason, the possibility spoken of becomes a certainty.
Reason tends to continuous advance; and its achieve-
ments are inherited by means of human culture, with
its special agency, human or rational speech, passing into
higher and more powerful developments in the form of
writing and again of *printing*. This is recognised in
Mr. Leslie Stephen's view of society; society is an
organic tissue, in virtue of the communion which exists
between its parts, through reason and through speech as
the embodiment of reason. The definition of civilisation
found in Professor Ritchie's *Darwinism and Politics*,
viz. "the sum of those contrivances which enable
human beings to advance independently of [biological]
heredity," points us in the right direction. Mr. Kidd
has missed the obvious truth because he is too intent on
biology, and too hurried in his glance at human society
and human reason. "Biologists" may prove if they
can "the non-transmission to offspring of qualities
acquired during the lifetime of the parent." If biologists
make out their case, they prove that such qualities are
not transmitted *biologically* or *organically*. They
cannot possibly show that "the effects of use and
education" are *not* "transmitted by inheritance."
Every time a child goes to school, he is entering upon
such an inheritance. True, he may inherit little "at
birth." What of that? Human progress cannot conceivably be regulated by "the accumulation of congenital
variations above the average" and by nothing else.

[1] Not that we can claim Darwin's authority for *this* belief.

That would imply that the world could gain nothing
from an intelligent sociologist unless he happened to
leave a son who was slightly more effective, socially,
than himself.   The truth is, that genius is rarely or
never reproduced in offspring, while yet progress is
secured by the *human*, the *rational methods.*   "The
sons" of the wise, as Old Testament language reminds
us, are other than his family after the flesh.   Even in
dying, "he shall see his seed."   Shakespeare is
Shakespeare, not to one generation merely, but to every
age.   Newton survives in the senior wranglers of to-day,
who could expose so many of his errors, and tell him so
many things he never dreamed of.   If Mr. Kidd's views
are solid, he has contributed directly to human evolution
by his very stimulating book; a contribution quite
independent of " accumulation of congenital variations";
while, if Mr. Kidd is wrong, one may hope to make
some small but direct contribution to human welfare by
exposing his fallacies.   Really it is almost ludicrous to
spend so much time in beating in an open door!   Yet
the conclusion pointed to is one of great scope and
importance, if we consider it thoughtfully.   Far from
being a mere accidental accretion upon the evolutionary
process, *reason has transformed everything.*   Reason is
not formal but constitutive.   Reason is not simply a
calculating machine, but a principle, whose workings are
seen both in nature and in man, both in knowledge and
in conduct.   It is not selfish, but moral behaviour that
deserves, and alone deserves, to be called rational.

And, if our view of reason changes, our view of
religion must change with it.   Religion is not the con-
tradiction but the fulfilment of reason.   For reason is
immanent in all things.   Every one of Mr. A. J.
Balfour's parallel pigeon-holes is a simple department

or manifestation of reason. " Ethics " and " Æsthetic " are as rational as abstract scientific knowledge ; how could they arise save in a rational consciousness ? And assuredly religion also must be a superstructure reared on the foundations of reason. But it is not true, as the intellectualists hold, that morals or æsthetics add nothing to that which is presented to us in knowledge. It is not true, as Hegelianism seems to imply, that goodness and beauty are mere allotropic forms of rational system, or that logic furnishes the master key to their meaning. Our knowledge is real knowledge, but has its limits ; and the meeting-point of these various stems lies underground, well out of sight. To God, their connection may be self-evident, their interdependence manifest ; to man, these great truths must continue largely matter of faith.

And therefore we do not speak idly when we say that reason finds its fulfilment not strictly in itself, but above and beyond itself, in religion. Men do not need religion to make it their interest to be good. That is, most deeply, our human interest. Yet man is in bondage. " The good that we approve we perform not ; the evil that we allow not, that we do." By a " pleasureless yielding " to " petty solicitations of circumstance," we destroy ourselves. Deliverance comes from above. " What the law could not do, in that it was weak through the flesh," has yet been done, and done in a Diviner way. Here is the true apologetic vindication of religion. Religion is no superfluity, though reason itself—so far as its influence goes—inclines us towards what is good. Religion is the breath of life, the touch of God, making that a reality, strong and victorious, which apart from it would be nothing but a faint aspiration or a bitter and hopeless regret.

CHAPTER XX

SUMMARY AND CONCLUSIONS

Self-contradictions—Comte is arbitrary—Biology has been reinforced by evolutionary theories, yielding different forms of sociological doctrine—1. Analogy, without struggle; Stephen—2. Continuity, without struggle; Spencer, Alexander (partly)—3. Analogy of Darwinism; Bagehot, Alexander, Ritchie (?)—4. Continuity of natural selection; Sutherland, Drummond (?), Kidd—None of these wholly succeed; old authorities will return !—Or idealism, which is compatible with the old authorities, may give us a more satisfying doctrine of evolution—What have we been taught ?—(1) A social organism exists—Idealism reinforces this lesson—(2) Struggle *has* been useful; *will* it not be ? as discussion ? as competition ?—In light of idealism this seems possible—Of fact, probable—Must not exaggerate its place; it is subordinate in life of reason—[Mallock]—Finally, does *progressiveness* of evolution make it a guide to conduct ?—Difficulties in *biology*; environment constant ?—Some forms have stopped !—Some never started !—Differentiation plainer here than progress—Reason makes for progress in *history*—Is it all-sufficient ? (Mill, Buckle)—Ancient civilisation failed—Morality and Christianity must safeguard modern civilisation.

AT the close of our wanderings, we propose to hold a stocktaking of the wisdom which we have picked up by the way. In other words, we shall run rapidly over the suggestions that have been brought before us, and try to estimate their value. We must note once again in how many voices and in how contradictory a fashion our teachers speak. Scientific sociology is still a hope rather than a fact; the " ethics of evolution " may mean any one of half-a-dozen or half-a-hundred things. The wisdom proffered to us is hydra-headed, it is million-tongued. But we must also try to decide, in general terms, what positive contribution to human guidance we may reasonably expect from "biological" inquiry. And we must look more closely at the definitions of evolution, especially at the question whether evolution is or is not identical in meaning with progress.

In Comte, the appeal to biology occupied a limited, almost a subordinate, position. Biology was the science next below sociology; it furnished the sociologist with suggestions; but decisive guidance was found in the wise man's inspection of human phenomena, or in his study of past history. We have seen, however, on how many distinct principles, and with how large an infusion of arbitrariness, Comte read off these lessons. In our opinion, such guidance as Comte yields was due to the working in him of the rational and moral nature of man. So far as biology in particular was of service, it gave him only parables.

Biology leaped into much greater prominence when the doctrine of organic evolution was propounded, and when evolution was further generalised (however vaguely) as a cosmic process. We distinguish two phases in this appeal—non-Darwinian evolution and Darwinian; and two forms of each, according as evolution is appealed to for *analogies* bearing on the social and ethical life of man, or according as an effort is made to *merge* that social and ethical life in a continuous evolution upon naturalistic lines.

First, we have evolution without the assertion of struggle applied to human affairs by way of analogy. This is chiefly exemplified in Mr. Stephen's doctrine of "social tissue," by which he serves himself heir to Comte. The doctrine, however, is without authority. It remains a hypothesis. We *may*, if we will, regard

morals as the laws of social welfare ; Mr. Stephen
would add, *versus* individual welfare.   No proof is
given that we *must* do so.

Again, part of Professor Alexander's theory falls
under this head, viz. the definition of goodness as
equilibrium.   Here a certain amount of proof is offered
us, viz. indirectly, in the form of hostile criticism of
rival naturalistic theories ; along with which we have
Mr. Alexander's assurance that the measure of truth
contained in idealistic ethics is incorporated in his own
formula.   We see no possible reason to forbid the
assertion that goodness is an equilibrium,—it is in the
farther working out of his views that Mr. Alexander
seems to compromise the interests of morality.   But
we remain unconvinced that " equilibrium " is either
the best or the only definition of moral excellence.

Secondly, we have evolution—still without vital
incorporation of the conception of struggle—in Mr.
Herbert Spencer, but now applied not simply by way
of " analogy " to the " social organism," but also—
and emphatically—to the whole cosmic process,[1] society
included.   At least, that is the effort of Mr. Spencer's
philosophy.   In its working out, as we noted, it falls
short of its aim, giving us rather a sequence of distinct
evolutions in different regions.   And for the guidance of
conduct Mr. Spencer does not keep steadily to the
suggestions furnished by cosmic evolution, but varies
his standpoints, and sets before us no fewer than three
ideals.

Thirdly, we have the Darwinian doctrine of struggle ;
and we take it for the moment as applied by way

[1] If Spencer is biological at all, it is in conceiving the universe itself
as an organism.   But that organism, by the definition, has no environment !

of analogy to human relations.   Now this Darwinian
doctrine is immensely important.   True, or false, or
half true—and we must not suppose that the truth of
evolution, even of organic evolution, stands or falls
with Darwinism—Darwinism still remains as when
first promulgated, the one dominant theory.   It
"holds the field."   While the factors of Spencer's
assumed cosmic evolution are shadowy and vague, the
factors of natural selection are—or seem to most minds
—plain and undeniable.   They may carry us far, or
they may carry us only a short distance; but they are
veræ causæ.

Darwinism is applied by Bagehot to nations and to
political life generally; by Professor Alexander to the
conflict of ethical ideals.   In neither case does the
assumed evolution follow the lines of true Darwinism.
Apart from war, Bagehot recognises imitation (cf. Professor Baldwin) and free discussion as the great factors
in progress or change.   Both of these are psychical
factors; they make for evolution directly, not indirectly; they may be expected to move much more
quickly than natural selection.   Professor Alexander
again (as we concluded), so far as he makes the conflict
of ethical ideals look like a Darwinian struggle, does
this by distorting his facts.   We may add here that his
vision of endlessly successive ideals has no authority
from Darwinism.   In nature, we see clearly that the
process of organic evolution has its definite limits,
and comes, now on one line and now on another, to a
fixed goal.   And the assertion that the reigning ideal
is *the* true ideal for its time, though only for its time,
finds no justification in the world of nature or in
Darwinism.   It implies some other philosophy; and
the unknown philosophy does not attract us.

Professor Ritchie is hard to group. He tells us that
Darwinism applies *mutatis mutandis* to human things.
" How else ? " With such a saving clause one might
predicate any attribute of any subject. The stuffed
horse of Wallenstein at Prague, with " only the head,
legs, and part of the body renewed," is the same
horse still, no doubt; *mutatis mutandis.* So long as
Professor Ritchie does not take a general view of the
changes which he recognises, we do not know whether
he believes in applying Darwinism by analogy to a
higher evolutionary region, or in extending Darwinism
to cover the whole field. Perhaps he has never faced
that distinction. In any case, his opinions are left too
vague to be estimated. He makes no attempt to find
guidance for conduct in Darwinism ; unless perhaps
from its " *not* sanctioning " struggle or *laissez faire ?*

Fourthly, however, we have the assertion of Darwinism as an all-embracing (organic and super-organic)
philosophy. This is found in Mr. A. Sutherland, and
we are not a little indebted to him for working it out
and showing where it leads. It means the denial of
the existence of human reason as a factor in the
cosmos, and of history as the embodiment of human
reason. This we might treat as reducing the position
ad absurdum. Against such extravagances not metaphysicians only protest, but evolutionists, like Darwin,[1]
Professor Karl Pearson, Professor Lloyd Morgan. They
have shown us in their capacity as men of science how
intelligence, as it arises in the animal world, limits,

---

[1] Darwin's denial of natural selection among the civilised is found in
Descent of Man, pp. 143, 618, quoted in Mr. K. Pearson's *Chances of
Death*, etc. i. pp. 127, 128. This may be set against the anti-ethical
suggestions of Darwin regarding bee-murder. While he was tempted to
interpret the higher by the lower in evolution, he was not pledged to
that error.

and finally banishes, natural selection. We have further seen that, while faithful to the conception of progress by elimination, Mr. Sutherland does not himself succeed in assuming the kind of elimination implied in true natural selection, viz. starvation or violent slaughter due to struggle.

Drummond did not definitely challenge natural selection. Probably he was a believer, and had no intention of excluding its operation from human society. He tried to show, mainly in the brute world, that it had limitations. The argument as he states it seems precarious, inadequate, and, in the light of a better philosophy, unnecessary.

We again find pure Darwinism, or rather pure natural selectionism—hyper-Darwinism, a Darwinism that goes beyond the master—asserted by Mr. Kidd following the lines of Weismann. We held his physiological basis to be insecure, and his sociological inferences illegitimate, even if it were possible to treat the problems of morality and sociology in an appendix to biology. But in point of fact *Social Evolution* turns as much upon the writer's private opinions regarding reason and religion as upon its view of struggle;[1] and that view, dissociating struggle from elimination, is not Darwin's view.

On the whole, then, this is what we have seen. The one attempt to give authority to biology as a guide for human conduct is the doctrine of evolution. The only accredited theory of naturalistic evolution is natural

---

[1] Professor Baldwin's *Social and Ethical Interpretations* furnishes a valuable criticism upon Mr. Kidd. Some of Mr. Baldwin's own positions seem obscure or questionable. But as he decisively subordinates the appeal to biology, he does not form part of the proper field of our present study.

selection. And it does not, it cannot, apply where reason is at work.

When this is more generally recognised we shall see a return of men's minds to the rejected authorities. Religion, conscience, philosophy, even intuitionalism, they will all come back, " trooping all together." Probably they will all have contributions to make to the social philosophy of the future. Faith in free-will must also return ; the ban of ostracism will be cancelled. Denial of freedom is exactly parallel to Mr. Sutherland's denial of reason, though many idealists have mixed themselves up with the one, while claiming to be champions of the other. But this is the truth ; there is a new factor distinguishing spirit from nature ; in knowledge it appears as reason, in conduct as will. One is delighted to find Professor Karl Pearson helping, though indirectly and involuntarily, to vindicate libertarianism.

Yet all is not done when we recognise the importance of reason and will. We are not at the end of social philosophy. We are only at the beginning of a better start. It was intolerable extravagance when Mr. Sutherland tried to make away with the existence or distinctive character of mind, though he only blurted out what many had been whispering behind their hands. And yet man has a body as well as a mind ; he has not ceased to be an animal, because he has become a spirit. He is still an organism. Probably old-fashioned ethics and libertarian philosophy made matters too easy for themselves by ignoring everything except the presence of reason and of free-will. We must keep both sides in view. May we advance a step further ? May we say that the two sides are not to be contemplated as two heterogeneous things—soul and body linked together like an ox and an ass yoked in the same team—but as naturally and necessarily related, or perhaps as in some deep sense identical? This is a programme hard to comprehend and hard to follow, but it has formed part of the noble endeavours of idealism. Idealism tells us that "such a being as man is, in such a world as the present," would not be more spiritual without his body. He is spiritual just because he is a human being—human body and human soul. Idealism holds that the animal functions, recognised in the life of man as "hunger and *love*," are no more anti-spiritual than spiritual, but rather the raw material of spirituality, of moral goodness, of character; life being the discipline and the ripening of character. It tells us that reason is the fulfilment (as well as the transformation) of nature; that man is the meaning, and therefore the goal, of the cosmic process which is seen in this world. What lover of humanity, what believer in its Divine goal, would refuse assent to this interpretation of man's place in the present world?

> Not soul helps flesh more now than flesh helps soul.

This is evolutionism, but a very different evolutionism from that studied in the previous pages. It would have been impossible therefore to try to bring in "Hegel" as well as "Darwin" in our present study. The new social philosophy, if it follows these lines, may be found to furnish not very much in the way of dogmatic sociology. It may well turn out that, on fuller reflection, the *a priori* scheme of "all possible societies" will shrink into very small compass, that the general programme formulated by wise teachers will be notably vague. That will not matter greatly. The wise social philosopher will not claim that the one fount of wisdom

for men or societies is the fountain which he has enclosed.
Ethics proper will be among his data. He will renounce
as fraudulent and absurd the attempt to deduce ethics
from schemes of physical or even of biological evolution.

Have we then learned nothing, it may be asked, from
the naturalistic schemes passed in review?

They have contradicted each other (and themselves)
so freely that it seems impossible to maintain they have
accomplished much. Nevertheless, we may notice their
two chief suggestions.

First, it has been suggested that society is an
organism; and Mr. Spencer, with difficulties to face
from the materialistic cast of his own philosophy (in its
spirit, if not in its letter), suggests that the universe is
an organism. These views will receive authoritative
support if we accept the idealist evolutionism. It will
no longer be a mere assertion, it will be part of a great
and subtle system of thought, if we now assert that
society is an organism; that its interests are paramount
to those of the individual; that in its good the individual
finds his own. Even the bold description of the universe
as an organism will be justified. The universe will be
revealed on deeper and fuller study as a system, not a
chance aggregation of disconnected parts, but a *cosmos*.
Chaos and chance will be banished to the region of bad
dreams. Reality will be viewed as the creation and the
image of thought. The relation between man and
nature will also be conceived as necessary or organic.
Everywhere will be traced such a priority of the whole
to the parts as organisms display to us. For the true
and beau-ideal organism is that which is more than an
organism, self-conscious reason.

Secondly, we cannot fail to observe a suggestion of a

different kind pressed upon us by the study of nature, the suggestion of the importance, nay more, of the indispensableness of struggle.   Of course, it is possible, or even probable, that the doctrine of natural selection is not the whole truth, even in the region of biology. Therefore it may be the case that the evolutionary study of nature, as conducted by our scientific leaders, hands on to sociology a stronger recommendation in favour of struggle than facts really warrant.   Further, we have agreed decidedly to repel the suggestion that natural selection strictly so-called has an appreciable effect in civilised society, or can account for advances in human morality.   Still, unless we utterly reject natural selection—perhaps one might even say, unless we close our eyes to manifest facts—we must admit that struggle exists in nature.   And it will need clear proof if we are to believe that the same necessity does not hold in human life.

Bagehot and Professor Alexander have mainly dwelt on the importance of free discussion.   That is a kind of competition.   It is very different, of course, from natural selection.   It implies reason and speech, and the possible wide diffusion of successful opinions,—a whole world of causes making for rapid advance in contrast to the heart-breaking tardiness of natural selection.   Still, it is a form of struggle.   And while defeat here points towards conversion rather than towards extinction, it would be absurd to say that defeat in argument is always painless. It is painful!   And it does not always make for progress. We have ceased to believe as confidently as the men of last generation in the immediate victory of truth.[1]   Yet

---

[1] There are interesting remarks on the evolution of beliefs in Dr. F. B. Jevons's *Introduction to the History of Religion* at the beginning of chap. xxvi.

if free discussion is maintained it will bring us in time
to the ultimate victory of truth; we still believe that.
And we have learned too that the refusal to give unbounded sway to argument is not wholly bad. It is not
pure perversity. It is partly due to the working of
deep but only half-articulate convictions and instincts.
Men cannot answer the glib logician, but they are sure
there is something upon their side of the case to which
he has failed to do justice. Socially and morally it
would be no advance if mankind laid aside their conservative misgivings, and sought to set up an age of
reason, with all the schoolboy enthusiasm of the Jacobins.
Convictions which are more slowly reached are more
deeply grounded.

Mr. Kidd lays stress upon the sort of competition
noted in political economy, personal competition between
man and man. Unquestionably this has been a vast
historical influence. It had its limits. Custom, as
economists since J. S. Mill have taught, very widely
forestalled competition in the history of human trade.
But the two factors are not necessarily inconsistent.
They may co-operate, as when custom fixes the amount
of a fee, while competition settles who shall do most
business and carry off most fees. In that way, or in
some fuller way, competition is likely to assert itself
irresistibly as the pressure intensifies. Struggle ensures
the maximum product.

But we have not done with custom when we have
recognised the increasing power of competition. In
other ways social custom has conditioned the working
of competition, notably in the *class standard of comfort.*
Men have never competed *en masse* for the necessaries
of life, or for the chance of piling up a fortune by miser-
liness. Both personal inclination and social pressure

have constrained those who rise in the world to modify their scale of expenses. Therefore the foolish prosperous man will tell the artisan that though richer he is no better off—not a bit—always on the wrong side of the account; and what to do with the boys—! A distribution of society into separate compartments tends thus to intensify struggle and to increase the total output.

The very fact that biology offers social science this second suggestion, in favour of struggle, shows in a crucial instance the unreliableness and self-contradictoriness of the biological lawgiving. If society is an organism, man ought to live for the general good. If struggle for existence is the true law of moral and social advance, then it is our duty to fight " for our own hands " with all our might. Which view is authoritative? Both cannot be; yet both are " the teaching of biology."

It may seem that any attempt to make room for struggle is equally inconsistent with that higher evolutionism based on reason, to which we have pointed. If reason promulgates a doctrine of the social organism, must not reason too feel nonplussed by the assertion that nature teaches the necessity of struggle? Yet, at the least, the philosopher's study of reason has prepared him to hear of an intenser struggle where conscious life prevails. He sees how self-consciousness draws a more definite line round the individual, making each organism a universe in itself, a microcosm, as no irrational creature is or could be. He perceives that the requirement sometimes addressed to man is foolishness,—that he should behave as a mere part in a larger social organism. It is idle to talk of such things. Self-consciousness puts an end to acquiescence in the mere suppression of the individual. But, if the first and lowest work of reason

is to break up the unity of sense, that unity may and
must be rebuilt in a higher fashion by the agencies of
morality and religion. So far we are willing to agree
with Mr. Kidd. Only we do not believe that the first
work of reason is its only work. We cannot admit that
morality and religion are divorced from reason.

Still, if it be true, as wise men taught long before
Darwin or Adam Smith, that life is a battle—if it be
true, as we have read in an old book, that the life of
a Christian man is a "fight of faith"—then we may
well expect to find conflict and struggle appearing as
elements in the orderliness and beneficence of the social
organism. Not indeed such struggle as is found in
natural selection; and very possibly not the "cut-
throat competition," as it is called, of unbridled
individualism, though in modern commerce we cut
prices, not throats, and nothing whatever is gained by
ignoring the advance which that fact implies. Not
every form of struggle, then, yet some form, and that a
keen one, is to be expected and desired. Morality still
leaves the individual personally *responsible*. He must
lead his own life, fight his own battle, gain his own prize.
And if, in the physical world, natural selection has
indeed been at work,—if, so far as it has been at
work, its cruel or seeming cruel methods have secured
this notable result, a teeming population of healthy,
vigorous creatures, fit in every fibre, fit or fittest on all
the varied lines along which evolution has moved, at all
the varied points which evolution has reached,—then
may it not be that social struggle, acting in union
doubtless with other forces, will give us an effective and
vigorous and truly happy human society? A man, or a
school, or a world is the better of hard work. And the
world will be kept hard at work; there is no throwing

off the yoke except for that unhappy minority, the idle classes.   Could we destroy social pressure we might find that we had simply destroyed the atmosphere which our souls breathe.

Yet, if we admit the permanence of struggle, we must strictly cross-examine the theories which are built on that fact, lest they exaggerate it.   They have called the process natural selection, in some cases, perhaps, because they were enamoured of struggle, and love-blinded to its dangers; in some cases but hardly in all cases.   What can be the reason why Darwinism has had so great a charm for many sociologists and moralists?

Perhaps the reason was that natural selection stated a method of progress without conscious known superintendence.   Many different forces struggled or competed—nature selected; environment selected; the struggle itself selected.   Many different patterns were aimed at; one pattern resulted, and no one had aimed at it.   Such at least is the suggestion underlying Mr. W. H. Mallock's definition of evolution as "the reasonable sequence of the unintended."[1]

But, if this be the meaning of the appeal to natural

[1] *Aristocracy and Evolution*, p. 97.   I merely observe how curiously the teleological suggestion recurs, even in a phrase which seems designed to exclude teleology.

Mr. Mallock's interesting book marks an advance, in so far as he insists that progress due to "great men" is more *rapid* than the physiological progress due to natural selection.   But he goes on to distinguish this advance, in the sphere of reason and realm of history, from mere biological evolution, on the ground that in the latter, wholes compete, while, in reason and history, parts of the social organism compete against each other.   That does not seem to hit the true line of difference, or to mark the real ground of the failure of biological sociology in the past, which Mr. Mallock once again deplores.   "Struggling parts" are not unknown in biological speculation.   Psychical progress, by great men or otherwise, is direct and therefore rapid.

Mr. Mallock overdoes his apotheosis of competition.   We will still believe that even the "great man" may rise to higher things than an exceptional hugeness of desire.

selection and to struggle, it almost forces us to ask whether our definition has gone deep enough. Are the competitors in reality so many distinct ultimate factors in progress? Or are they all held in the grasp of one great evolving system? *not*, however, to be defined as matter and motion growing more complex! Is the relation between the different forces simply or mainly one of rivalry; is it not predominantly one of co-operation? Is history · a Kilkenny cat struggle between nations, or in history is struggle itself subordinated to an evolution of mankind? Ought an enlightened nation to regard its neighbours mainly as rivals, or mainly as brothers in the common tasks of civilisation? And so with ethical conceptions; is the history of moral thought mainly a struggle of system against system, of ideal against ideal, or is it an evolution of one ideal? And is each moralist pledged by fidelity to his own views to eat up and destroy his rivals, or may he also be the conscious servant of a wider truth? Even in nature, one more and more questions the adequacy of the view which regards the various organisms simply as each other's rivals, the co-operating forces simply as happening to coincide. And, when we pass on to the fuller " symbiosis " of reason and morality, the Darwinian formulæ snap in two. Men superficially regarded are competitors, but essentially they are their brothers' keepers, and members of one great fellowship.

Yet one more attempt may be made to find a guide for conduct in phenomenal knowledge, if evolution everywhere and necessarily is equivalent to progress. We have met this view before—more than once; first in the appeal to history, then in Mr. Spencer's cosmic doctrine of evolution. Here too, if anywhere, the contendings of Mr. C. W. Williams [1] are relevant. Though it offers very little guidance in detail, yet this assertion demands to be looked at. It can be held, and is, apart from any claim to knowledge of the factors of evolution.

We do not attempt to say anything further regarding merely physical evolution. In spite of Mr. Spencer, we doubt the possibility of laying down laws *a priori* for that process. But we must consider, in the first place, biological evolution, or the evolution of species. And secondly, we shall pass on to speak of evolution in human history.

If we might assume natural selection to be the key to organic evolution, we should have a good deal of reason for identifying evolution with progress. "Natural selection" seems to imply the transforming of minute random variations into definite serviceable changes. If everywhere there is movement, the movement ought everywhere to result in progressive efficiency or adaptedness. Yet the assertion is a difficult one.

First of all, there is one very plain condition, which presumably no critic will question, but which ought to be made explicit. If evolution is to mean progress, it must at least imply continuous adjustment to a constant environment. If the environment changes, if there is no continuity in the definition of "fitness," there can be no real progress. Dissatisfied with my dwelling, I build myself a house exactly suited to my personal needs. That is a real improvement. But forthwith I have to accept an appointment in a different town, and must sell my new house at a loss for whatever it will fetch. The improvement due to building for myself is forfeited, and turns to the opposite. Now in the far-off

[1] *Review of the Systems of Ethics founded on Evolution.*

past our planet is said to have passed through more
than one ice age.   Of course so tremendous a change in
environmental conditions involved the forfeiting of past
progress.   The tests were all (however gradually)
altered.   The last became first, and the first last.   The
unfit were now found fit, while the fit proved unfit.
Physiological capital was fatally depreciated, like
machinery thrown out of use by a better invention.
Only here there was no better invention.   There was no
continuous progress.   There was discontinuity and a
change of conditions.   Evolution then will scarcely
mean progress unless *first* it is continuous evolution.
But continuity in evolution of species implies constancy
of environment.   No doubt, speaking broadly, we have
had such continuity on the earth for a good many
æons.

Secondly, a difficulty occurs as to those species
which seem unchanged from remote geological times.
Drummond's *Ascent of Man* has been the one of our
authorities which has told us most about these.   There
are shells, it seems, absolutely unchanged through
many ages, because they had "arrived."   They had
reached the limit of possible development on the line
which they had chosen.   More important still is the
case of man, whose physiological improvement, according to Fiske, has been superseded and arrested by the
emergence of reason, and whose cranial development,
according to Professor Cleland, has gone about as far
as is possible under the laws of space in their bearing
on the constitution of the human body.   We cannot
therefore say—in spite of all Darwinising moralists—
that "everything is in flux," moving "from change to
change eternally."   Evolution seems to be a definitely
limited movement, exhausting its possibilities, now in

one direction, now in another, now in some low forms
of organised life and again in the highest. Further,
was this evolution exactly identical with progress even
while it lasted? In the case of man, we shall assume
that it was; was it equally so in the case of the shells?
Progress means advance on one line; evolution seems
to mean radiation in many directions. It may be
taken then as meaning differentiation; or the gradual
filling out, by mechanical process, of a designed and
purposed scheme; or the eliciting of all the possibilities
latent in "protoplasm" at the first. Of these conflict-
ing interpretations the first might suggest Spencer; the
second, a Christian teleology; the third, Spinozistic
Pantheism.

There seems no doubt that origin of species by
natural selection would imply variation, or differentiation of race from race. Animal A preys upon animal B,
and threatens to exterminate it. Several specimens
of B may deal with the difficulty in several distinct
fashions. The swift B will run away from A and
make its escape. The cunning B will hide itself from
A and elude notice. The strong B will stand up to
A manfully, and, after a few struggles, will teach A
to seek his prey by preference among less warlike
creatures. There is no one means of survival in
the struggle; there are several. At any time, for
any species, there are innumerable possible advantages.
Candidates for nature's examination can and do
specialise. It seems, therefore, that fitnesses are produced, but fitnesses of manifold types. Progressive
improvement (given constancy of environment) everywhere results, but it results upon different lines, and
the clearest outcome of the process is the transition
from the monotony of a few types to an almost infinite

variety. Of course we must remember that variation in other types constitutes a change in the "environment" of any one type, whether the altered neighbour was a former competitor, or a former ally, or liable formerly to be preyed upon, or making prey formerly of the type in question. It follows that a constant environment, such as "progress" involves, can only be affirmed in a relative and limited sense. And therefore we must similarly qualify the connected assertion of continuous organic advance and improvement as the result of natural selection.

A third difficulty strikes one in connection with the lowest organisms. Certain shells or the human physique have ceased progressing because they have reached the allotted goal; good, but why have the lowest not moved up? Experimental science refuses to admit abiogenesis. Wherever life came from at first, it does not now arise from a rearrangement of dead matter. If "all were in motion," including the initiation *de novo* of life, then we should see through the difficulty. Infusorians would be infusorians—only that and nothing more—because they had not had time to climb up the ladder. But apparently, in point of fact, they have had just as much time as the cedars of Lebanon or the crowning race of man; and in that time, of course, a vastly greater number of generations. Then why are they still mere common infusorians? Take it either way; why have they not progressed out of that state of being; or, at any rate, why have they not varied? Through billions on billions of generations—to put it modestly—they have been competing against each other and against the cruelty of environment. Why are they still no fitter? or, if they are fit enough to survive—why has any other organism taken the trouble to build up new and higher forms of life? There seems reason to think that this consideration points to some grave flaw or gap in naturalistic theories of evolution.[1]

On the whole, from our human point of view, we consider that the evolution of species has been attended with progress, because "higher" animals and plants have appeared, and, above all, because man has emerged. We must also admit that the evolutionary process has been attended with a vast differentiation of life into forms not all of them admirable from an æsthetic or from a *quasi*-moral point of view. Whether there is advance upon each divergent line, as differentiation takes place, may appear doubtful, though the theory seems to affirm it. Differentiation appears to be proclaimed far more clearly than progress, alike by the theory of natural selection and by the phenomena of living but irrational nature.

When we turn to human evolution, we find at once that there are changes. The law of differentiation has still been at work, though its conditions are obscure and ill-comprehended. We have negroes, Esquimaux, Mongols, Caucasians, all probably of the same stock, all very dissimilar. Yet even here there is something quite different from animal evolution. Races of men do not dwell simply side by side, indifferent to each other, as plant and animal races do. You may, of course, have a society built in separate compartments, as in the institution of caste, or in the simpler

[1] Mr. A. R. Wallace suggests that the lower types fill up the few places *of that kind* which nature allots! Mr. Wallace is a little inclined to switch on and off selective struggle at his arbitrary pleasure and convenience. His own position is exceptional (see p. 210); but, on the naturalistic view, *ought* not the lowest forms to be *originating* before our eyes?

and more familiar case of slavery. Yet this differentiation, gross and excessive as it is, belongs to another
region of things from animal differentiation. The
many castes—or the slaves and the oppressors—constitute together one society. The potential unity of
the race, implied in reason, has already that notable
consequence. Accordingly, the marked physiological
differentiation of the various races of mankind does
not seem to have taken place in a society having
relations even of neighbourhood between its several
parts. It has been guessed that race differentiation
was due to natural selection in different regions of the
world, those naturally superior to cold surviving within
the Arctic circle, and those who enjoyed immunity
from fever surviving in the tropics. At any rate, the
differentiating process came first. While man was
mainly an animal—or (what is nearly the same thing)
while men were divided from their fellows by geo-
graphical barriers—they diverged physiologically; and
no doubt they also diverged socially. But, as soon
as reason began to assert itself and make its way, the
tendency to differentiation was held in check by a
tendency to unity—a growing unity of culture and
custom pointing to an ultimate far-off unity of the
whole race. The different branches of the human
stock can borrow from each other as kindred tribes of
animals cannot do. Even if, for a time, the aristocratic
few have no mind to help the ignorant many, yet the
ignorant many are eager to copy the envied few.
Simple survival of the fittest and neglect of the unfit
is never long the rule in human affairs. Levelling up
is one of the earliest manifestations of reason, when set
free to do its work.

In the first instance, as between different societies,

this process no doubt takes place through war. The stronger race conquers, and *the defeated race eagerly imitates the conquerors*. This would be fatal to progress if an inferior race were capable of mastering higher races on the field of battle. But, as Bagehot has forcibly pointed out, up to a certain distance the opposite is true; through many ages, we may be sure that the best man or best race will win at the game of war. Yet how different are the consequences from those of a merely animal victory! Instead of stubbornly clinging to their old ways, the conquered usually develop an enthusiasm for their conquerors. Like the natives of America, they regard the higher race as half Divine beings. A whole civilisation or semi-civilisation falls into wreck, and a higher or stronger one takes its place. It is truly pitiful to read of some of the forms this takes, *e.g.* in Rhodesian Africa, where the black women despise and desert the men of their own tribe, and know nothing better than to yield themselves to the white men.

Later on in evolution a race may be conquered which is possessed of high attainments in culture. But by this time the higher culture is able to rise superior to the rude test of efficiency on the field of battle, and the great task of unifying humanity still goes on, though under somewhat different conditions. Greek culture poured eastward like a flood in the track of Alexander's conquests, but it filtered westwards too in spite of the arms of Metellus or Mummius. *Græcia capta*—the thing has become a proverb. Not less notable and not less hackneyed is the case of the barbarian conquerors of the Roman empire, who went to school to the civilisation which they had overrun. Even the break-up of the empire into many national kingdoms, and the dis-

appearance of the common Latin speech before the new
Romance formations or the native languages of Teutonic
races,—even these changes did not signify mere retrogression.   The new nations were not indifferent to the
rest of Christendom.   They felt themselves members of
one great civilisation, making their characteristic contributions to the common stock, and making them all
the better because each nation took its own way.   Even
the aberrations of modern nationalism do not imply any
forsaking of this standpoint.   The nation or the race is
determined to be its own untrammelled self; yet it is
willing, nay it claims, to be one of the great family of
civilised mankind.   The civilised world moves essentially as a whole.   What one race gains, all share.   Is
it not plain that our posterity will come to make the
same assertion regarding the whole of mankind?   Ultimately even the most backward races must join the
fellowship.   Ultimately even the least philanthropic
must share the burden of the weak.   "We without
them cannot be made perfect."

Human evolution then differs from evolution in the
organic world.   It does not mean progressive divergence
of type from type, but progressive unifying, all differentiation being strictly held subordinate to the unity
prescribed by reason.

Does human evolution then mean progress?   Assuredly man can frame the conception of progress, and
once he has done so, nothing will satisfy him save steady
progressive advance and improvement.

Reason grasps this conception, and reason itself, or
the free development of intelligence, is certainly one
condition of historic human progress.   Without reason
there can be no movement onwards or upwards at the
more rapid pace at which history moves.   Very likely

Bagehot's explanation is true (so far as it goes) that reason was first emancipated among those races which "happened" to have free political constitutions, and acquired in politics the instinct of free inquiry. The further question, what maintains progress? or what leads to new advance? needs no discussion. We need not, like Professor Ritchie, seek biological analogies, or look to the mixture of races[1] as the cause of new "varieties." Once the spring is opened up, it flows. There is in intelligence, freely exercised and firmly organised, a constant tendency towards improvement. This is no metaphysical assumption like Mr. Herbert Spencer's evolutionary doctrine; it is plain fact that where the reason of man is at work, a force has come into operation which makes for progress by an internal law.

Is that force absolutely sufficient? Does it carry with it all the allied forces of our nature so far as other forces are distinguishable from it? That is the doctrine laid down by Mill, and more explicitly affirmed over against the claims of morality by Buckle.[2] From criminal statistics Buckle drew the extraordinarily sweeping inference that goodness and sin were fixed quantities, and that intelligence was the varying and progressive factor in human nature. As well might he have watched half-a-dozen waves break on the beach, and then announced that the tide was neither ebbing nor flowing. Moral progress, no doubt, is slow in comparison with material progress; but who will dare to affirm that in a world of evolution goodness alone fails to evolve?

---

[1] Compare Bagehot as above; also Dr. Tiele's Gifford Lectures.

[2] It must be remembered that Mill and Buckle were pre-Darwinian writers or thinkers. They had no opportunity of asking themselves, Does reason alter the working of evolution? The working of evolution was not among their data.

When we transport this question into the field of history, we are struck with the phenomenon of the breakdown of ancient civilisation. The defeat of the Roman Empire as a fighting force was the least of its failures. Intellectually, too, it was exhausted; it was the transmitter rather than the possessor and enjoyer of the great classical culture. The barbarian inroads, Sir Henry Maine tells us, may have saved Europe from the fate of China. Intellect was enfeebled, and morality, as in all protracted civilisations hitherto, had suffered deep perversion. What will guarantee us against a recurrence of such failure? A recurrence would be decisive. There are no unspoiled barbarian races to take up the torch once more and carry it onwards.

Now there are two advantages on the side of the modern world. We have a better method in physical science, and we have a better religion, or the religion we share with the Christianised empire is better acclimatised in our soil. Either the intellectual or the moral revival; either the Renaissance or the Reformation. *In hoc signo vincemus.*

Physical science is no doubt a great and a lasting boon. Discoveries large and small are made, and will be made; they pay so well. Bacon was right in his enthusiastic eulogies on the " fruitfulness " of the science which he dimly foresaw. But that is hardly the question. Even without much physical science the humane culture of the great ancient world had vast powers for intellectual progress. In spite of this it broke down. Can science as applied to physical nature really guarantee the world against moral paralysis?

Others will hold with Mr. Lecky that the decisive factors in progress are *moral*, and—not perhaps with Mr. Lecky—that in Christianity, or, as Christians prefer to say, in Jesus Christ, and in Him alone, we have the pledge of the human world's fulfilling its destiny, of the vanquishing of all the obstacles that can arise, of the great career's reaching, at last, that

one far-off divine event
To which the whole creation moves.

INDEX

Printed by R. & R. CLARK, LIMITED, *Edinburgh.*